Fisheries Access for Alaska— Charting the Future: Workshop Proceedings

January 12-13, 2016
Anchorage, Alaska

PAULA CULLENBERG, EDITOR

WORKSHOP PLANNING TEAM
Paula Cullenberg
Linda Behnken
Courtney Carothers
Edward Davis
Rachel Donkersloot
Duncan Fields
Nicole Kimball
Rep. Jonathan Kreiss-Tomkins
Norman Van Vactor
Bob Waldrop

AK-SG-16-02
$10.00
Published by Alaska Sea Grant
University of Alaska Fairbanks

for Meredith, free!

Library of Congress Cataloging-in-Publication Data

Names: Cullenberg, Paula, editor. | Alaska Sea Grant College Program.
Title: Fisheries access for Alaska--charting the future : workshop proceedings, january 12-13, 2016, Anchorage, Alaska / Paula Cullenberg, editor.
Description: Fairbanks : Alaska Sea Grant, University of Alaska Fairbanks, 2016.
Identifiers: LCCN 2016029209 | ISBN 9781566121866 (alk. paper)
Subjects: LCSH: Fisheries--Limited entry licenses--Alaska--Congresses. | Fishery policy--Alaska--Citizen participation--Congresses.
Classification: LCC SH329.L53 F57 2016 | DDC 333.95/61709798--dc23
LC record available at https://lccn.loc.gov/2016029209
ISBN 978-1-56612-186-6
doi: 10.4027/faacfwp.2016

Citation

Cullenberg, P., ed. 2016. Fisheries Access for Alaska—Charting the Future: Workshop Proceedings. Alaska Sea Grant, University of Alaska Fairbanks, AK-SG-16-02, Fairbanks. http://doi.org/10.4027/faacfwp.2016

Credits

Alaska Sea Grant is supported by the US Department of Commerce, NOAA National Sea Grant, grant NA14OAR4170079 (A/152-32) and by the University of Alaska Fairbanks with state funds. Sea Grant is a partnership with public and private sectors combining research, education, and extension. This national network of universities meets changing environmental and economic needs of people in coastal, ocean, and Great Lakes regions. Cover photo by Deborah Mercy.

Alaska Sea Grant
University of Alaska Fairbanks
Fairbanks, AK 99775-5040
(888) 789-0090
Fax (907) 474-6285
alaskaseagrant.org

Table of Contents

Introduction

Access by residents to local commercial fishing resources is critical to a healthy fishing community. Many of Alaska's fishing communities have seen a decline in the number of limited entry fishing permits and amount of harvest quota held by local residents. In addition, entry of younger Alaskans into commercial fishing permit or quota ownership has lagged—in 1975 the average age of transferable permit holders was 42.7 and by 2013 it was 49.7. In some rural communities, fishing access is down to a few individuals and the entry rate by young people is minimal.

In January 2016 Alaska Sea Grant and a group of committed partners held a two-day workshop designed to better understand the decline of local fishing access and its impact on Alaska's communities and economy, and to explore options to reverse the trend. Over 100 individuals participated, affiliated with Alaska's fishing associations and communities, Alaska Native organizations and tribes, state and federal regulators, legal advisors, academics, and other policy makers. Participants included experts from Alaska, the US East Coast, Europe, and Iceland.

Workshop speakers and participants addressed trends in fishing access privileges held by Alaskans, defined the scope of the problem, and suggested guidance statements for Alaska policy makers to consider and potentially adopt. Legal and regulatory constraints, financial constraints, and other roadblocks were outlined. Speakers and attendees presented examples of possible solutions, such as permit banks, community-based fishing trusts, educational permits, apprenticeships, and local small-scale community fisheries. Potential pathways to achieve outcomes were laid down at various levels—legislative, regulatory, municipal, tribal, and business entrepreneurship or financing.

The workshop spurred discussion statewide about creating viable policy solutions to increase access to fisheries for Alaskans. In March 2016, workshop steering committee members presented concerns and solutions to the Alaska Legislature House Bush Caucus in Juneau, Alaska.

These workshop proceedings serve to document the discussions and presentations at Fisheries Access for Alaska—Charting the Future. The content of the presentations and discussions are captured in this volume and on the Alaska Sea Grant website.

Paula Cullenberg, Director
Alaska Sea Grant
July 2016

(Deborah Mercy photo.)

Let's Move the Dial on Fishing Access

Alaska Lt. Governor Byron Mallott
Juneau, Alaska

Speech excerpts: The lack of fisheries access by Alaskans is one of the most important public policy matters to pursue today. Access to fisheries is vital to the strength of our communities. We need to achieve success in this.

I grew up in Yakutat and then went to boarding school. At the school, the kids from Yakutat always had some cash because they fished. Since then things have changed. Hoonah used to have a beautiful fishing fleet, but now has only two boats. People need work that they can take pride in. Fishing is a life opportunity that makes sense and is good for children to get started in. It was the best for us in Yakutat. Back then we took it for granted, but today we no longer have that opportunity for families and communities.

When change came from limited entry, it was very quick. It was mostly done in a political process. There was no discussion about the long-term future, and there were unintended consequences. The value of the fishery has been monetized—it became a function of the marketplace.

So we are here today. Fishing permits that are wanted by individual fishermen in the communities remain with the industrial fishers. Some Alaska coastal communities still have hard scars from the canned seafood industry, a time when individual fishermen also wanted ownership. Now permits go where the money is. That is the way capitalism works.

Today Alaska is in a serious fiscal circumstance—we face consequences of actions taken years ago. The Alaska State Legislature and administration have to make the right choices to prevent a catastrophe.

We have two Alaskas—urban Alaska and rural Alaska. Fishing is a key opportunity in rural Alaska. The marketplace cannot be the only determiner for who gets a permit. Governor Walker and I are behind efforts of this workshop. We want to help you "move the dial" on the fisheries access issue.

Lieutenant Governor Byron Mallott addresses workshop participants. (Deborah Mercy photo.)

Comments on Improving Alaska Fisheries Access

Sam Cotten, Commissioner
Alaska Department of Fish and Game, Juneau, Alaska

Thank you very much for inviting me to participate. Thanks for your work on this. Your group and agenda are very impressive.

I have a few things to offer on the subject. There has been a decline in access rights held by coastal residents, especially those who live in small rural communities. We all know that people have to have access. There are few options for economic activity without it. Without that access the fishing community sustainability is also seriously threatened. And we are guided some by the constitution. We are supposed to make resources available for maximum use consistent with the public interest. Public interest certainly includes ensuring access. And the utilization of our resources for maximum benefit of its people also guides us on the fact that fishing access is one of the highest priorities for the Alaska Department of Fish and Game.

Providing access is one of our more important functions. This is something we have been working on with legislators, especially with Representative Jonathan Kreiss-Tomkins on some of his legislative initiatives. We will continue to do that.

I am very much looking forward to any recommendations that you may have for improvements to the current decline. While we search for means of improvement we also want to make sure that we avoid further decline. And that's where I want to talk about the federal arena. The North Pacific Fishery Management Council manages fisheries outside of three miles but we have to acknowledge that the boats tie up in Alaska boat harbors, the fish are processed in Alaska fishing communities, and certainly directly affect the people who live and work in coastal Alaska.

Proposals are currently being advanced that would assign exclusive access privileges to historic participants. I say further because on your agenda I noticed there was some discussion about the federal limited entry program, LLP (License Limitation Program) as it is known. Those permits aren't cheap but these new proposals would assign quota.

There are arguments in favor of this that have to do with efficiency. There is some merit to the arguments, but the result would be additional restrictions and additional expense to do with access to these fisheries. Currently the proposals are just with trawl fisheries, but many of these efficiency arguments would also be applicable to longline, pot gear, and jig fishermen. They've been a source of concern for me and for many people who are dealing with this issue.

Proponents suggest that in order to avoid bycatch you have to rationalize, and you have to use additional tools to stay under the current limits. We've offered alternatives that would provide individual accountability, and that would not result in additional restrictions or additional expense. There is not a lot of agreement in the trawl fleet. It is an ongoing discussion at the Council. We are pretty strong—we do not want to make it more difficult or more expensive for people of Alaska to get into those fisheries.

The bulk of the Bering Sea is owned by people who don't live in Alaska except the CDQ (Community Development Quota) groups, which has been a very successful program. It has done great things for the economies of a lot of the western Alaska communities that would not have had a chance at such an economy otherwise. The Gulf of Alaska is different—there is a lot of local participation. It is growing, and recent efforts by the Board of Fish can increase opportunities by adding more fish to state water fisheries. I think we are in a position to make some gains but we want to make sure we don't lose ground. We don't want to head in the wrong direction.

Thank you again. Let us know if there are things we can do to further this cause.

Perspectives on the Significance of Access to Fisheries for Alaska Residents and Communities

Steve J. Langdon
University of Alaska Anchorage, Anchorage, Alaska

The fisheries resources have been and are of enormous significance to Alaska residents and communities for literally thousands of years. I have had the good fortune to be a student of, and participant in, fisheries themselves and policies associated with them for over 40 years.

I commercial fished from 1973 to 1976 in Southeast Alaska. I was involved in purse seining with Native captains in Craig, Alaska; handlining for halibut and hand trolling for salmon, taught by local Tlingits; and beach seining with Tlingits for subsistence. I lived and observed the values and culture of fisheries and its significance to communities, residents, and generations. This formed the basis of my doctoral dissertation in anthropology.

The following points lead to a broad perspective on fisheries access for Alaska residents and communities.

- Who benefits from fisheries in Alaska?
- What were the circumstances?
- What happened?
- What have been the outcomes?
- What is at stake?
- What are the possibilities?
- Crises and radical realignments—"shock doctrine." [1]

1. *The Shock Doctrine* is a 2007 book by Naomi Klein in which she presents the argument that crises of various kinds have been used to construct property rights regimes that empower the few and the rich and marginalize the many and the poor. The downstream effects of these transformations, she argues, have reduced the quality of life for many and dramatically limited their opportunities for betterment.

- Institutions should be designed to sustain values and practices of those they affect.

Contexts of limiting access: state

The impetus for statehood was driven by the imperative to ban fish traps. The absentee ownership and economic benefit from floating fish traps first introduced in the 1910s were primarily realized by the corporate elite living in Seattle. Over 50% of the harvest was taken by the traps and the Alaska resident fishermen deeply resented this condition. The intent of statehood supporters was to ban traps so that control of resources and their benefits would be realized by Alaska resident fishermen using their gear. This entire movement created the archetype of the "independent boat owning fisherman" as the Alaska counterpart to the Western cowboy as hero. Following statehood in 1959, banning traps was the first action of the new government on which the people voted (see ballot below). But banning fish traps led canneries to shift investment to vessels leading to a rapid expansion of the fleet, especially from Washington state.

Section 1. Ballot
Ordinance no. 3, Abolition of Fish Traps

Each elector who offers to vote upon the ratification of the constitution may, upon the same ballot, vote on a third proposition, which shall be as follows: "Shall Ordinance Number Three of the Alaska constitutional convention, prohibiting the use of fish traps for the taking of salmon for commercial purposes in the coastal waters of the State, be adopted?"

Yes______
No ______

Section 2. Effect of Referendum

If the constitution shall be adopted by the electors and if a majority of all the votes cast for and against this ordinance favor its adoption, then the following shall become operative upon the effective date of the constitution: "As a matter of immediate public necessity, to relieve economic distress among individual fishermen and those dependent upon them for a livelihood, to conserve the rapidly dwindling supply of salmon in Alaska, to insure fair competition among those engaged in commercial fishing, and to make manifest the will of the people of Alaska, the use of fish traps for the taking of salmon for commercial purposes is hereby prohibited in all the coastal waters of the State."

Adopting the state constitution and outlawing fish traps

The 1974 Boldt Decision upholding Indian treaty rights of 50% of harvestable salmon was the major impetus of the Washington purse seine fleet relocating to Alaska. In the late 1960s major climatic cooling led to salmon declines. The lowest point in Alaska was in the 1970s when Boldt Decision impacts were also felt. In 1968 legislation that required Alaska residency for commercial fishing licenses was deemed unconstitutional, also having an impact.

Alaska State Constitution
Article 8. Natural Resources, Section 2. General Authority

The legislature shall provide for the utilization, development, and conservation of all natural resources belonging to the State, including land and waters, for the maximum benefit of its people.

Alaska State Constitution
Article I: Declaration of Rights, Section 23. Resident Preference

This constitution does not prohibit the State from granting preferences, on the basis of Alaska residence, to residents of the State over nonresidents to the extent permitted by the Constitution of the United States.

Crisis in Alaska salmon fisheries

A deepening crisis of biology and economy began with sharply reduced runs in the late 1960s. Initially, the legislature passed a law for an Alaska-only fishery but that was deemed unconstitutional. The new program, built on market transferability of permits, was developed by 1973 and led to the passage of the Alaska limited entry program with permits to be held by individuals, the return of the heroic independent fisherman. This required a constitutional amendment and the Alaska Supreme Court ultimately upheld its constitutionality. The first permits were issued in 1975. Individual ownership imposed structural constraints and there were implementation problems.

- The fisheries in Bristol Bay closed for one year in 1975 due to the weakness of the runs.
- Native fisheries experienced a profound cultural disconnect due to rules adopted for awarding points based on vessel ownership and not gear ownership.

- Assignment of a permit to one of the partners, especially in rural western Alaska villages, disrupted families, relationships, and practices.
- Many who were part of the Alaska Native "baby boom," which offset the national baby boom by about 10 years, were left out of the permit assignment.
- Permits were non-collateralizable except to the State for loans.

Legislative study of transfer patterns

It was evident by the late 1970s that permits were migrating out of Native hands, especially in villages. In 1978 I approached House majority leader Nels Anderson about conducting a study of permit transfers. He supported the idea but told me that it must not use ethnicity as a variable. The regional districts for permitting commercial fishermen in gear types became the basis for limited entry permit awards and therefore it was decided that would be the best way to go. I designed a system based on the regional areas and their communities to see how permits were initially awarded among various categories of residents including nonresidents. Tracking of permit transfers from 1975 to 1979 showed general declines in rural ownership with dramatic decline in certain fisheries. Factors associated with selling and buying were examined. Fundamental cultural disconnects were identified for rural Native populations. Salmon purse seine fishing used to be the foundation of Southeast Native village economy, society, and culture, but it had been dramatically reduced by 1979 due to permit losses and now it has all but disappeared.

The study found that rural permit holders often had less than optimum gear, so their earnings were below the fishery average. Thus they were often desperate due to lack of earnings and alternatives. Rural permit holders had the dilemma of whom to transfer the permit to among many sons and other family members, and they had little information about how to deal with permit bureaucracy. Loan programs were deeply tilted toward residents with education levels, collateral, loan histories, and familiarity with bureaucratic institutions not found among Alaska Natives. State loans had virtually no rural Native participants.

Addressing some problems

As a result of the study, a target loan program was created for rural areas in 1980. But later a Commercial Fisheries Entry Commission (CFEC) study claimed State-subsidized and target loans were driving up permit prices, and the programs were stopped.

Institutional permit ownership was not contemplated at the time. Tribal ownership was first suggested in 1984 by myself with an eye toward tribes being authorized to hold permits to prevent further declines in holdings in smaller communities.

When Perry Eaton, of Kodiak, was head of Alaska Village Initiatives, he asked me to do an evaluation of causes of permit loss in Bristol Bay and to suggest some possible steps to take to stem the tide. As a result, several programs were explored in an effort to provide information among Bristol Bay owners and those Bristol Bay residents seeking permits to make transfers possible among the regional cohort. A Permit Information Clearinghouse was developed, with the intent to provide information on permit availability to regional residents.

Native corporation loans, another possible step to help resident fishermen, are not feasible because a permit is not collateral except to the State. If a corporation is not paid back, the money is gone and possibly the permit. The individuals seeking the loans rarely have the collateral necessary to cover the loan amount since the permit could not be used as collateral.

Contexts of limiting access: federal

The halibut and sablefish fishery limited access program is different due to the structure of the resource and its management, and the fact that they are under federal jurisdiction. The appearance of crisis in the halibut fishery was due to different factors. Continuous entry of fishermen increased effort leading to shorter and shorter seasons. The layoff program for rotating fishing effort among vessels was abandoned. Annual harvest quotas allowed for the division of catch based on fishing history. Individual fishing quotas were the new management technology but again followed the heroic model of owner-operator and denied any consideration of awarding rights through quota to crewmen who were instrumental in the conduct of halibut and black cod fisheries.

A study on halibut participants demonstrated there was a large cohort of crewmen who relied on the fishery, many of whom were professional. The individual transferable quota (ITQ) proposal was extremely controversial. ITQ was supported by large vessel Seattle-based owners whose shares in the fishery were declining. ITQ was opposed by Alaskans, for the most part, who were increasing their share of the fishery at the time.

I served on the North Pacific Fishery Management Council Scientific and Statistical Committee at the time of consideration in 1985. We said it would cut out the crewmen from a share of quota, but were told not to worry, that State loan programs would help Alaskans cope. The Knowles Administration opposed ITQs precisely due to the findings of the transfer pattern studies—it could harm Alaska interests, especially rural.

ITQ impacts and new initiatives

As predicted, the same trajectory of rural Native ITQ loss followed quickly, as had occurred in limited entry. Between 1995 and 1999, village permit ownership across the Gulf of Alaska plummeted. Federal policies due to the Magnuson-Stevens Act were open to the possibility of rectifying inequities and damaging impacts. A National Academy of Sciences study on ITQs, "Sharing the Fish" (1999), made a strong case for designing institutions to accommodate different values, to insure the viability of fishing communities and also to treat crewmen fairly.

A Community Quota Program was proposed to the Gulf of Alaska Coastal Community Coalition, which was then worked into a proposal that moved very rapidly through federal processes. Gulf of Alaska villages quickly moved to create local Community Quota Entities (CQEs), where nonprofits need to buy permits. Over 30 villages created them, but only three villages have been successful in using them to enhance local fishing opportunities. Share prices and the availability of quota, funds, and financial considerations such as loan terms and length of loan, have prevented more widespread use of this program.

Small-scale, local, community-based fisheries

In rural Alaska communities kinship, local knowledge, heritage, and transgenerational ties are key. For example, in Yakutat local subsistence and commercial salmon fishing are integrated. Sharing subsistence salmon and commercial with elders is fundamental.

Programmatic considerations

In 2016 the state's crises are extreme, with immense budget shortfalls, declining employment opportunities, community stress, and bleak futures. Fisheries opportunities can be a bright spot. Efforts must be multifaceted, and coordinated if possible. Institutions need to be responsive and flexible—tribes are the most important institutions in most Alaska Native villages. We must abandon ideological stances to meet circumstances and expand opportunity rather than limit it. Crises produce opportunities—the great institutions that created the US middle class arose out of the depths of the recession. Make this crisis one that works for the most people, not the most capital.

Village Historical Perspective on Limited Entry and Federal Limited Access

Freddie Christiansen and Emil Christiansen
Old Harbor Native Corporation, Anchorage, Alaska

Freddie Christiansen: We were born on Kodiak island and we are from Old Harbor, one of six villages on the island. We have spent our lifetimes experiencing the negative impacts of limited entry permits and halibut individual fishing quotas (IFQs). We have lifelong experiences of observing what occurred in the village. Growing up we did not have much in the village. We had a lot of love and care, but no toys or bicycles. Out our back door, the beach and the mountains were our playground.

I recall when I was about eight or nine years old a State of Alaska representative came to the village to convince the elders to sign a petition supporting limited entry fishing. All of our elders signed the petition to allow it to go forward. But when it did finally pass some villagers didn't get their limited entry permit, for example because they didn't make the qualifying years—1967-1971. Some elders who signed the petition retired from fishing in 1965-1966, after 30 years; they did not receive permits. I knew from that point on that it wasn't going to be a good program for our people and communities, and especially the villages.

I recall playing with six or seven friends on a beach in Old Harbor, with small wooden seiners carved by our elders. Of those seven friends I am the only one who is a fishing captain today. All the rest were locked out of the fisheries and went on to do other things. Three are no longer with us—they died by age 35. The limited entry program had a very negative impact on them, by not allowing them to fulfill their dream to become a skipper on a fishing boat.

As a kid I identified some things with the limited entry program and individual fishing quota (IFQ). Before IFQs a neighbor would go out on "derby days" fishing in four halibut openers a year, and catch 1,000-1,500 pounds. He was supporting his wife, two children, and son-in-law.

His catch of 1,500 pounds per opener added up to 6-10,000 pounds per year. For salmon they got $1.50–$2.00 a pound. It is important to note that some years they had bad salmon seasons, and it was the halibut that got them through the winter. Well, after he got his IFQ share of less than 1,000 pounds, there was no way he was able to survive on that.

These are things I have observed and watched over the years. I think there is a need for the state or federal side to be more considerate and do more consultation with the village when they are passing fishing laws. I am opposed to further regulations that exclude communities from participating in fisheries, such as Gulf of Alaska rationalization. We have noticed that the Community Development Quota (CDQ) program has been successful, and different parts of the state should be allowed to have such a program.

Emil Christiansen: I can give testimony to fight for the villages. I'm 63 years old, and I've had my day. Today I'm here to advocate for the future of the children. The State of Alaska should listen to the problems we have in harvesting our resources. We have three great resources in Alaska—oil, fisheries, and timber. At the state level they have not paid attention to the fishery because they had a lot of money from oil. All of a sudden today we are asking—where's the fishery gone? It's gone out of state and it continues to go.

In the village we were raised to expect to become a fisherman. We thought there would always be fishing. I used to fish Tanner from January to March-April. I fished herring in April-June. I fished salmon in June-September, and king crab in September-October. All that is gone now, taken away from our kids because of limited entry. I always thought that the right way to do it would be—if you come from out of state, you buy a permit to go fishing, and then you leave. You don't own that permit. Then the village kids could continue to get into the fishery. But the limited entry permit system got too expensive—it just isn't right for villages. In the village there are very few jobs available—you are a fisherman, a school teacher, a janitor, or a postmaster.

I see the CDQ program working in the Bering Sea, and I think we could do that in the Gulf of Alaska. If we want to continue to save the fisheries in the state we have to work together, not fight among ourselves. I'm not here asking to take anything away from anyone. I just want to make sure our children have a piece of it. I think this is what it's all about. I've had my fun fishing. The children want to do the same thing but the opportunity is not there. I expressed what I think. This group sitting here—if we work together we can fix this. Thank you.

Changes in the Distribution of Commercial Fisheries Entry Permits

Marcus Gho
Alaska Commercial Fisheries Entry Commission, Juneau, Alaska

Alaska State Constitution

Alaska entered statehood in 1959 during a period when salmon stocks were struggling due to overfishing. Recognizing the importance of Alaska's natural resources, the framers of the state constitution included an article that defined how natural resources are to be utilized in Alaska. Article VIII, sections 1 and 3 of the state constitution read in part, "It is the policy of the State to encourage the ... development of its resources by making them available for maximum use consistent with the public interest. ... Wherever occurring in their natural state, fish, wildlife, and waters are reserved to the people for common use."

Establishing the Commercial Fisheries Entry Commission

By the late 1950s, the salmon resource was in a decline, which resulted in less income for commercial fishermen. At the same time, more and more participants entered the fishery. When Alaska became a state, management of fisheries such as salmon were transferred to the State of Alaska.

The Alaska Department of Fish and Game strove, with some success, to employ management strategies that allowed for long term sustainability of Alaska's fishery resources. However, their efforts were not viewed as sufficient to ensure that commercial fishing Alaskans could continue to enjoy a fishing way of life and continued economic benefits. In 1972, the Alaskan people amended the state constitution allowing for license limitation. License limitation in Alaska is a system that requires permits to fish, and in fisheries such as salmon, allows only a certain number of permits to ensure both sustainability of the fish population and preservation of economic health of the fishery. The following year, the Limited Entry Act passed, establishing the Commercial Fisheries Entry Commission, also known as CFEC.

CFEC is a result of the legislative action—we do not make the law but are a manifestation of the initial law, subsequent changes, and over four decades of litigation. CFEC is a standalone entity with regulatory authority separated from our administratively attached Department of Fish and Game. CFEC has just over two dozen employees not including vacant positions. By statute, employees include three commissioners with supporting administrative and legal staff in addition to four units: adjudications, licensing, information technology, and research staff.

Limitation of permit types

If an individual in Alaska would like privilege to legally commercial fish in Alaska, he or she will need a CFEC permit. While the majority of permits are not limited, over the years many have been. In 1975 the first limitation of 19 salmon fisheries occurred (Table 1). The last limitations occurred in 2004, bringing the total number of limited fisheries to 78.

Table 1. Limitation of permit types.

Dates	Number of permit types	Cumulative total
1975	19	19
1976	6	25
1977-78	4	29
1980-87	16	45
1988-91	6	51
1997	7	58
1998	10	68
1999-2002	8	76
2004	2	78

When fisheries are limited, only a specific set of permits is issued, the majority of which are transferable. Transferable permits allow family members to maintain access to traditional fisheries when permit holders choose to keep the permits in the family. The Alaska legislature felt this was important, therefore the transferable feature of permits was selected among alternative reallocation approaches. Permits are issued to individuals and are required to be present during fishing activities, and leasing of permits is illegal, both of which ensure that fishing activities are not controlled by absentee permit holders.

Permits are issued for a specific species, gear type, and geographic region. In some cases, the permits specify along with their type, a certain level of fishing capacity. Initially, individuals were allowed to hold only a single permit for each permit type. This law was amended in 2002, now allowing an individual to hold two salmon permits; however, the second permit would not have any additional fishing privileges. In 2006, the Alaska legislature modified the law again to allow the Alaska Board of Fisheries to determine what fishing privileges the second permit could provide.

Changes in the distribution of Alaska's Commercial Fisheries Entry Permits

The CFEC Research Unit frequently receives requests to query our data sets. One place to find data to better understand who is participating in Alaska's limited fisheries is the CFEC annual publication, *Changes in the Distribution of Alaska's Commercial Fisheries Entry Permits.* This publication is more commonly known as the Transfer Study. The Transfer Study has been produced most years since 1981. Throughout the years this document continues to evolve with additional chapters, downloadable Excel tables, and other changes to make the data more accessible. Generally speaking, data in this publication and paper refer to cumulative results; however, the Transfer Study contains a wealth of data presented in time series as well.

I will also present some data from the CFEC Alaskan City tables, which can be downloaded as csv files and compiled, which I did for this presentation. While the CFEC Alaskan City tables also include ex-vessel earnings and fishery participation, my focus today is on the number of permits held by permit holders in various Alaska communities.

Data sources

Fishing participation requires an abundant amount of recordkeeping by the Alaska Department of Fish and Game. CFEC conducts data sharing with ADF&G, which means that CFEC has a wealth of information to draw from to evaluate fishery management outcomes. These data, along with the CFEC proprietary permit file and the federal census data, are used to generate the tables of the Transfer Study and the online CFEC Alaskan City tables.

The most recent edition of the Transfer Study includes 38 tables and covers data from 1975 to 2014, has additional tables in the appendices that fill more than 300 pages, and has an executive summary. The Transfer Study only includes adjudicated permanent permits.

The CFEC Alaskan City tables include landing and permit holding data for 326 Alaska communities as well as nonresidents for the years spanning 1980 to 2014. The tables include permanent permits as well as interim-entry, moratorium, and open-access permits. Permit holdings are counted only at year end. Note that these figures include only permanent permits and do not count emergency transfer permits. Both the Transfer Study and the CFEC Alaskan City tables will be updated with preliminary 2015 catch data by late spring.

Residency

The CFEC Research Unit is frequently asked to provide statistics relating to the residency of permit holders. The five most common subcategories used to describe residency status of permit holders include combinations of Rural/Urban, Local/Nonlocal, and nonresident. For example, an Alaska Rural Nonlocal would be a permit holder from Nuiqsut who fishes in Bristol Bay, and an Alaska Urban Nonlocal might be somebody from Anchorage who fishes in Ketchikan.

Depending on the research project, we sometimes narrow down our definition to Alaska/nonresident. It is easier to simplify by combining groups.

Summary of annual net changes in statewide permit holdings

The residency distribution of permits may change by three means: permits may be transferred from one person to another; permit holders may change their domicile; or permits may be cancelled or removed by CFEC.

CFEC is commonly asked questions concerning the distribution of permits held by Alaska Rural Locals. Generally, there has been an outflow of permits from Alaska Rural Locals. The black line in the figures depicts the sum of all **net** changes each year since 1975. The overall net change in the distribution of Alaska Rural Local permits from 1975 through 2014 is a decline of 2,304 permits, representing 28% (2,304/8,245 x 100 = 27.9%) of the permits originally issued to this group. As mentioned earlier, this includes the **net** effects of transfers, migrations, and cancellations.

Net transfers

Transferred permits contribute to changes in holdings by Alaska Rural Locals. The line with squares represents net changes to Alaska Rural Local permit holdings due to transfers. Transferred permits include permits which are sold, inherited, or given away; transfers between family members are frequently gifted. From the late 1970s until the late 1990s many permits were transferred out of the hands of Alaska Rural Locals, but beginning with the late 1990s the net transfers of permits shifted toward Alaska Rural Locals.

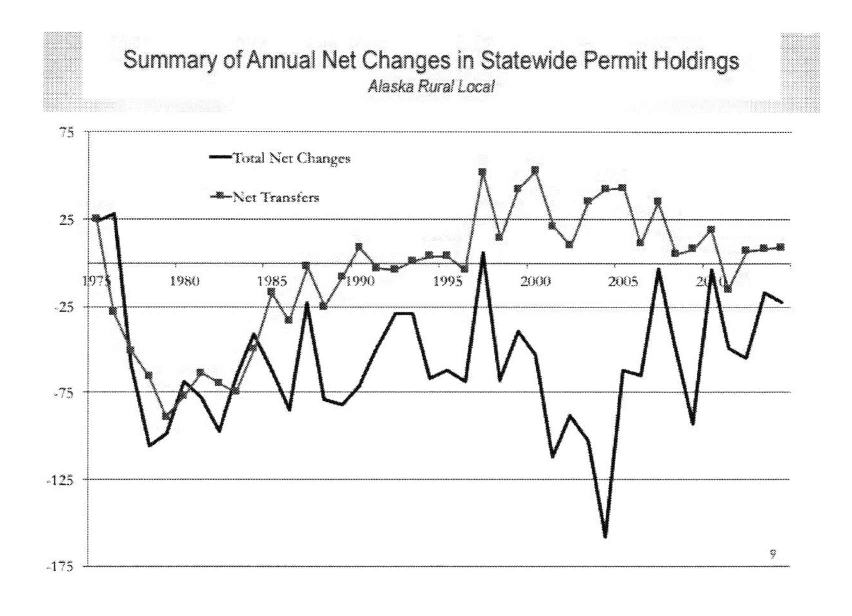

Net migrations

Another way permit distribution may change is through migration. Migration of permit holders occurs when permit holders change their domicile to a different community. An example would be when a Bristol Bay permit holder from Togiak moves to Anchorage; the status of their permit would change from Alaska Rural Local to Alaska Urban Nonlocal. As shown, migration has had a substantial effect on Alaska Rural Local permits with a total net loss of permits in the majority of years since 1975. Overall, migration has had a greater influence on the change in distribution of permits held by Alaska Rural Locals than transfers have.

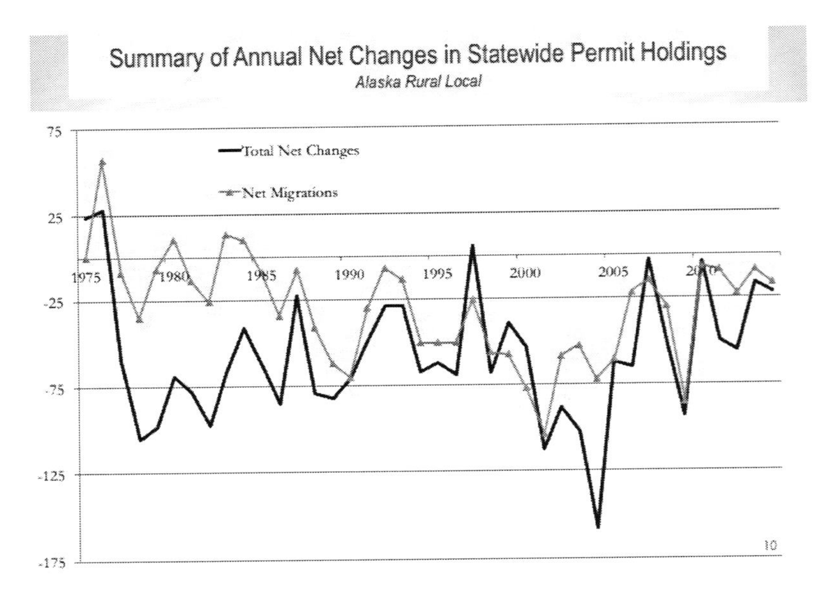

Net cancellations

The third redistribution of permits from Alaska Rural Local status is administrative in nature. Canceled permits typically occur on permits that were issued as non-transferable; these permits are relinquished upon the death of the permit holder or when the permit holder decides not to renew the permit. Small numbers of other cancellations occur due to criminal actions, administrative or judicial procedures, or voluntary relinquishments such as when there is a permit buyout program. Cancellations of renewable permits are most often offset by reinstated permits in the same year of cancellation with only a very small number transferring in the subsequent year. In 1980 the hand troll fishery was limited, and a large proportion of the permits were issued as non-transferable. In the early 2000s, many of these permits were cancelled as the permit holders retired from fishing. Almost half of all cancelled permits were non-transferable hand troll permits.

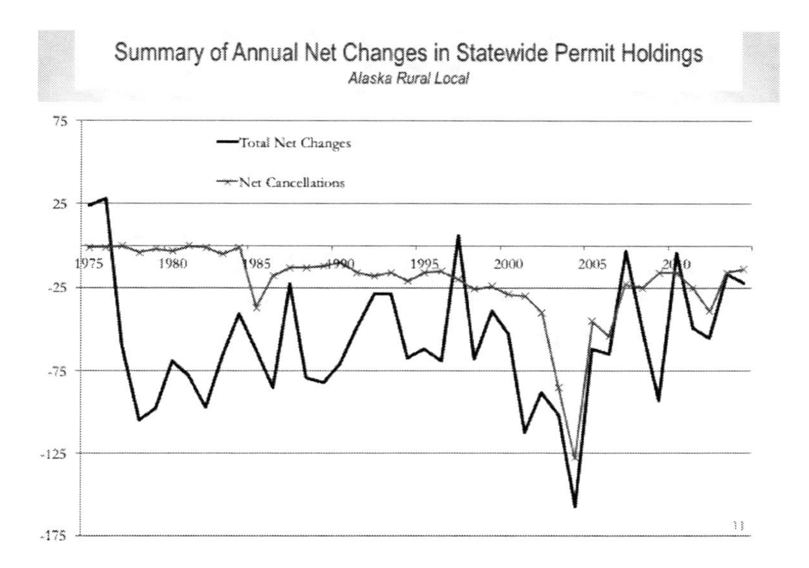

Locally held rural permits

Rates of change for individual permit types vary from the overall totals. The graph shows movements of Alaska Rural Local permit holders due to transfers in two of the salmon permits. Locally held rural permits in the Bristol Bay drift gillnet permit type saw an exodus due to permit sales most notably in the 1970s and 1980s. Southeast Alaska power troll fishery permit holders have generally transferred permits to rural locals.

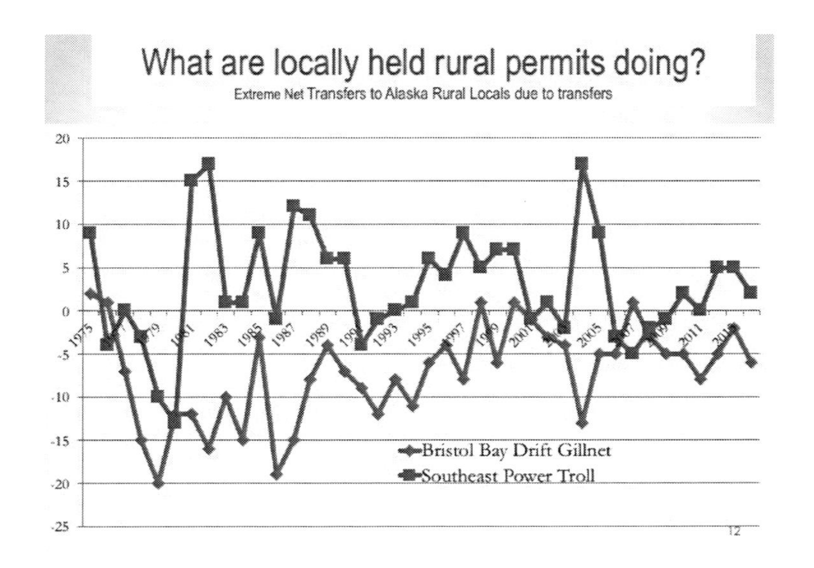

Aging of permit holders

The Transfer Study provides statistics on the age of permit holders. The black line in the graphs depicts the mean age of all limited CFEC permit holders from 1975 to present. If no permit holders transferred away their permits and none of them died, then the average age would have naturally incremented by one each year. If permit holders transferred their permits to individuals older than themselves, the average annual increase would be greater than one. As permits are transferred to a younger generation, the average age decreases proportionately to the age gap of the transfer recipients. The black line takes all three actions into account.

It is often more interesting to look at the extremes. The Prince William Sound herring pound and Cook Inlet herring purse seine permit holder average age has seen a constant increase. Both of these permit types have very few transfers.

Although not as common, other fisheries have an average decrease in the age of permit holders such as for the Southeast king/tanner crab permit and the Prince William Sound salmon set gillnet permit. While it is interesting to consider the aging of permit holders, it is important to keep in mind that the average age of working class individuals has also increased over the same time period.

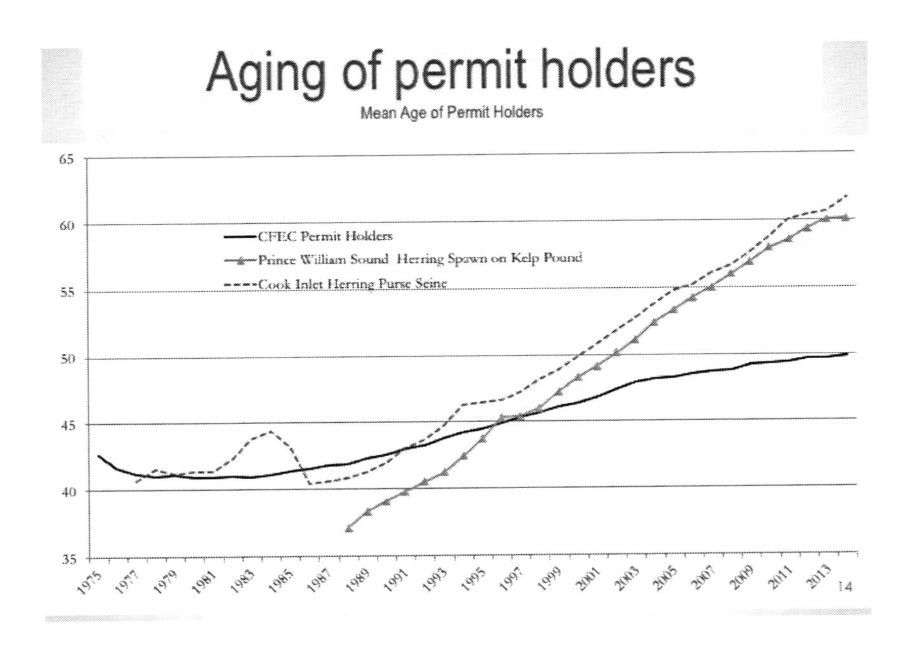

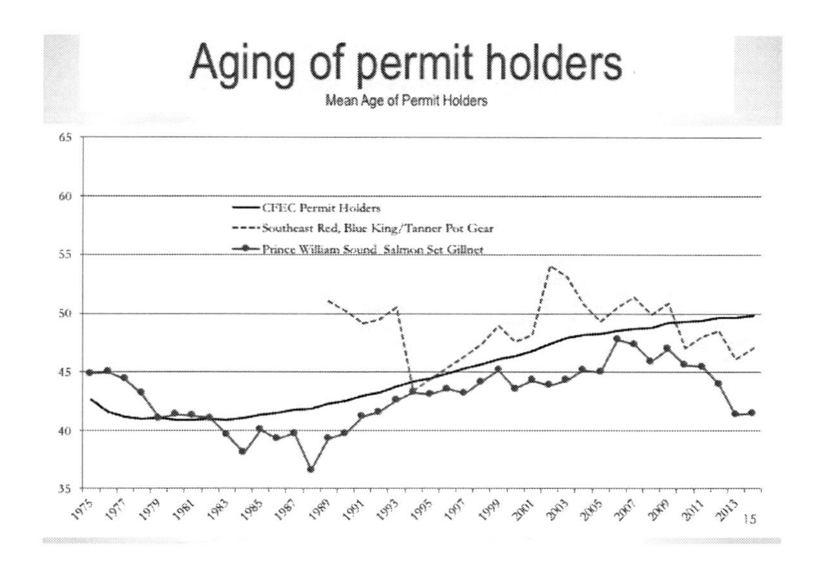

Relationships of transferees

When permit holders transfer their permits, who are the permits going to? Data from the Transfer Study can be depicted longitudinally in terms of transfer recipients. When CFEC was crafted by the Alaska legislature, the crafters wanted to allow fishing people the ability to transfer permits to other family members to support the concept of fishing families; therefore transfers could occur as gifts or inheritances. Since 1980, CFEC has required a survey to be completed for each transfer. These surveys provide information such as the relationship of the transferor and recipient, and the value placed on permit sales, the reasons for transfer. About two in five transfers are between family members for all permits transferred since 1980. The Bristol Bay salmon drift permit has the highest count of transactions, roughly one in seven of all transfers, with distributions comparable to statewide distributions. Some permit types differ. The Southeast sea cucumber diving gear permit has a high number of transfer recipients in the "Other" category (no a priori relationship). The Lower Yukon herring gillnet permit is of a completely different culture: almost every single transfer was to an immediate family member.

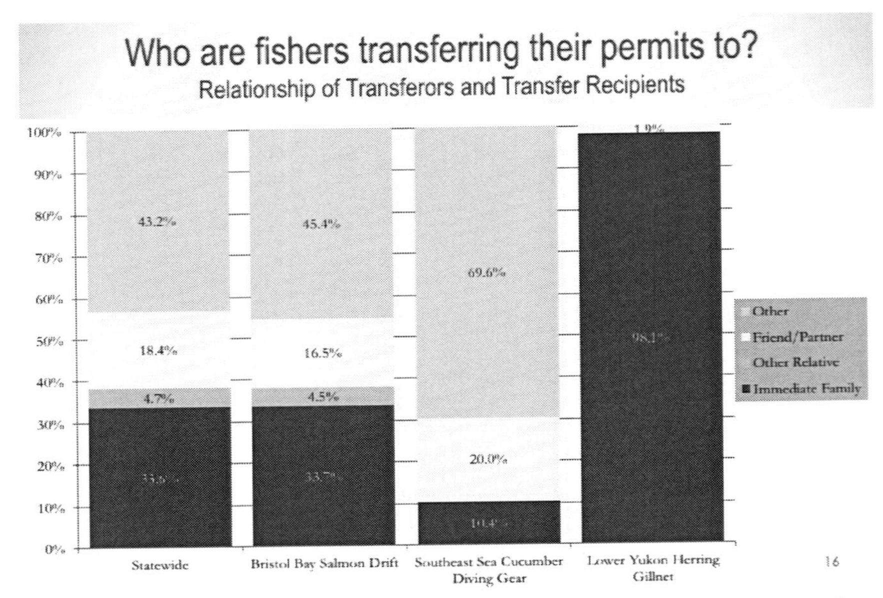

Immediate family: Statewide 33.6%. BB salmon drift 33.7%. SE sea cucumber 10.4%. Lower Yukon herring 98.1%.

Multiple permit holdings

In 2002, statutory changes were made that allow individuals to hold a second permit. In 2006, the law was further amended to allow the Alaska Board of Fisheries to decide what fishing privileges could be had with a second permit. Since then, the Board of Fisheries has exercised that right and allowed permit stacking, the ability to use an additional complement of gear with a second permit in a few set gillnet fisheries.

Communities going to/from 0 permits

The following information is from the CFEC Alaskan City tables. One inadequacy of tables is that they count only the number of permits as they increase or decrease without considering how the general population of the community has changed.

Tatitlek had 30 permits and Mount Edgecumbe had 116 total permits in 1975, but by the end of 2014 neither had any commercial fishing permits. Neither Nikiski and Nikolaevsk had permit holders living in their communities in1975, but by the end of 2014 had 49 and 37 permits respectively.

The same type of criteria can be applied by looking only at salmon permits. In Nikiski, there were 0 permits held in 1975, but by the end of 2014 there were 42 salmon permits. Mount Edgecumbe residents went from 54 salmon permits down to 0 permits over the same time period. I did not look at the type of salmon permits, but I should mention that CFEC allows individuals as young as 10 to be able to hold set gillnet permits. For all other permit types, permit holders must be at least 16.

Largest changes in permit holdings

Another view of changes in permit holdings has to do with communities having fewer permits. Anchorage and Dillingham have experienced the largest decline in total permits, with net losses of 2,132 and 571 respectively. Petersburg and Sitka have seen increases by 175 and 166 total permits.

One way in which permit counts were artificially higher in the earlier years is by permit adjudication during the year. In some instances, permit holders had both a permanent permit and their interim-use permit the same year and these can be counted twice. Another reason for a decline in permits is that when a fishery is limited, not all participants from the previously open access fishery qualify or even apply for a permanent permit.

Both Juneau and Ketchikan have seen a net decline of 319 permits, while Homer has experienced an increase of 331 permits and Wasilla has seen an increase of 141 permits. Note that these counts are for permits, and not permit holders. Individuals may hold multiple gear/location types of permits. Individuals are also allowed to hold two salmon permits, and in some instances the second permit provides for an additional component of gear.

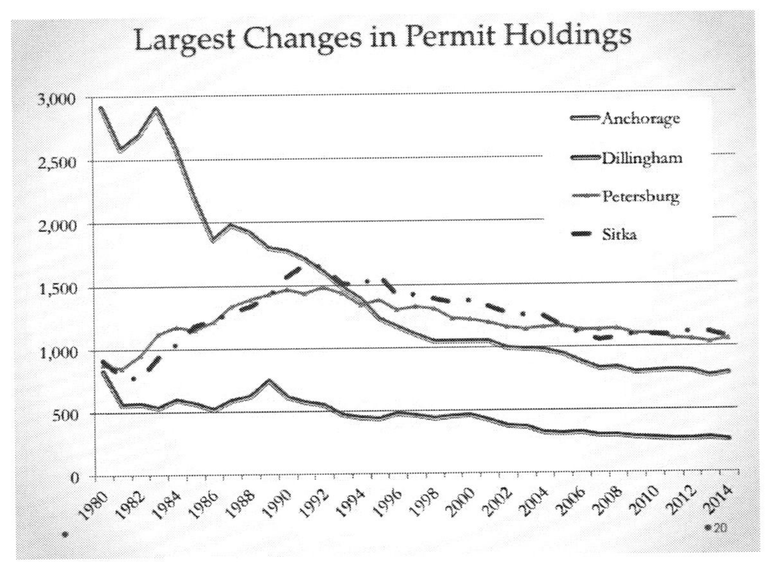

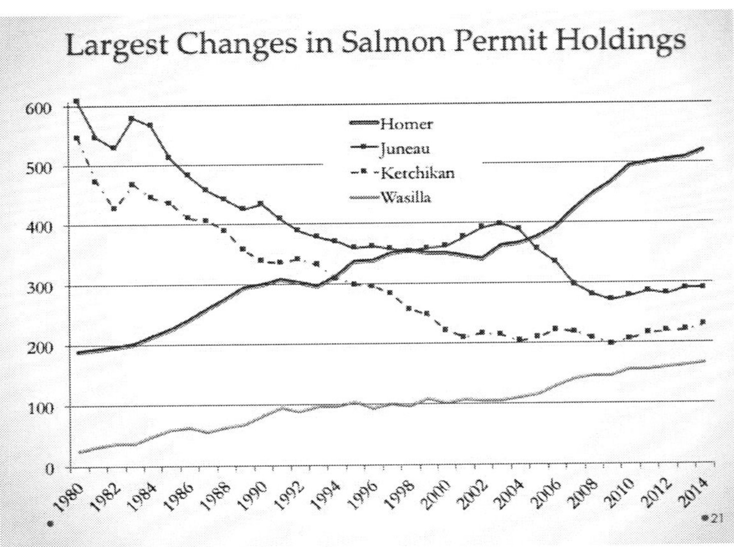

Conclusion

There have been many changes in the distribution of who holds fishing privileges. In some communities there have been increases, and in others there have been decreases. The structure of limited entry allows for the transferability of permits from individuals to individuals and does not restrict where he or she lives. This has resulted in changes over the years that have affected communities and fisheries differently.

Economics of Limited Entry Permit Transfers, Migration and Local Ownership

Gunnar Knapp
Institute of Social and Economic Research, University of Alaska Anchorage, Anchorage, Alaska, Gunnar.Knapp@uaa.alaska.edu

Introduction

I have been interested in questions about the economics of limited entry permit transfers, migration and local ownership for a long time.

A general question:

- What determines the extent to which transferable fishery access rights are owned[1] by local residents?

More specific questions for Alaska's limited entry salmon fisheries:

- What explains past change in the local share of permits?
- How do management policies affect local permit ownership?
- What can be done to slow or reverse local permit loss?

I wrote a paper five years ago (Knapp, G. 2011. Local permit ownership in Alaska salmon fisheries. Marine Policy 35:658-666). But since then I've been too busy with other work to study this issue more.

Over time, the share of access rights owned by "local" residents may change, through two mechanisms:

- **Net transfers** of rights from local residents to nonlocal residents.
- **Net migration** of rights-holders out of the region

Does local ownership matter? Economists have generally shown relatively little interest in whether assets are "locally" owned. Economists tend to think transfers or migration that reduce local ownership are "good" (or at least not "bad") because:

1. I use the terms "own" and "ownership" for convenience even though holders of these rights don't necessarily legally "own" them.

- Transfers allocate permits to those who value them most.
- Migration allows permit holders to live where they can derive the most value from permits.

This perspective underlies why we generally don't restrict or ban transfers, and don't restrict or ban migration by permit holders. In contrast to economists, other social scientists (for example, anthropologists) and local residents often argue that whether permits are locally owned **is** important because it affects where fish are landed and processed, where vessels are home-ported, where fishing income is spent, and where fishing crew are hired. Local ownership also affects whether young people have opportunities to learn about fishing, and whether fishing communities remain fishing communities.

I think local ownership of fishery access rights does matter. My goal is to understand how economic factors affect local ownership over time.

Take-away points

Any time we create fishery access rights without restrictions on transferability or where the rights holders live, over time very powerful market forces may affect where the rights holders live.

If we allocate the access rights initially by some nonmarket method (e.g., past fishery participation or residency) over time the ownership pattern will tend to shift toward what a market initial allocation (e.g., auctioning them off) would have resulted in.

It may be very difficult to overcome these market forces as long as the access rights are transferable and not subject to restrictions on where the rights-holders live.

One option is to create access rights that are not transferable or that are subject to restrictions on where the rights-holders live. But there are other issues associated with these types of access rights.

Trends in permit transfers, migration, and local ownership

There is a lot of variation between Alaska salmon fisheries in the extent to which local permit holdings have declined:

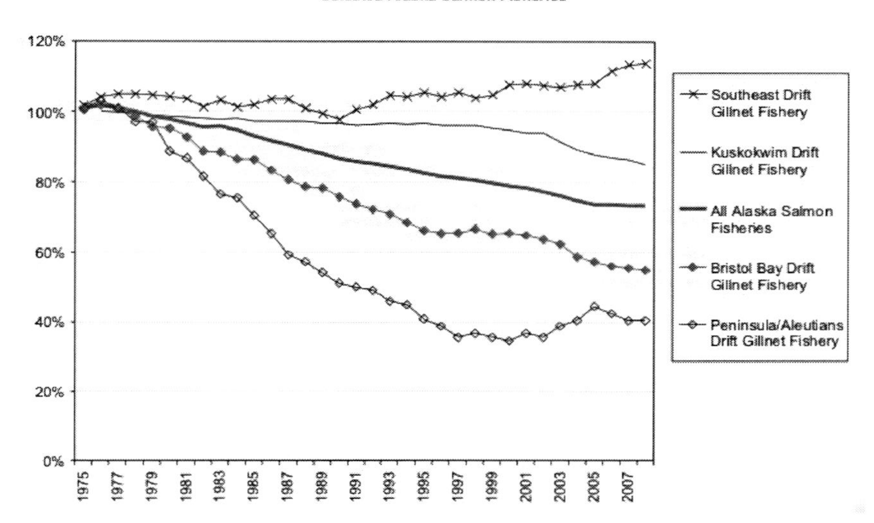

There is also a lot of variation between Alaska salmon fisheries in the extent to which net transfers have contributed to a decline in local permit holdings:

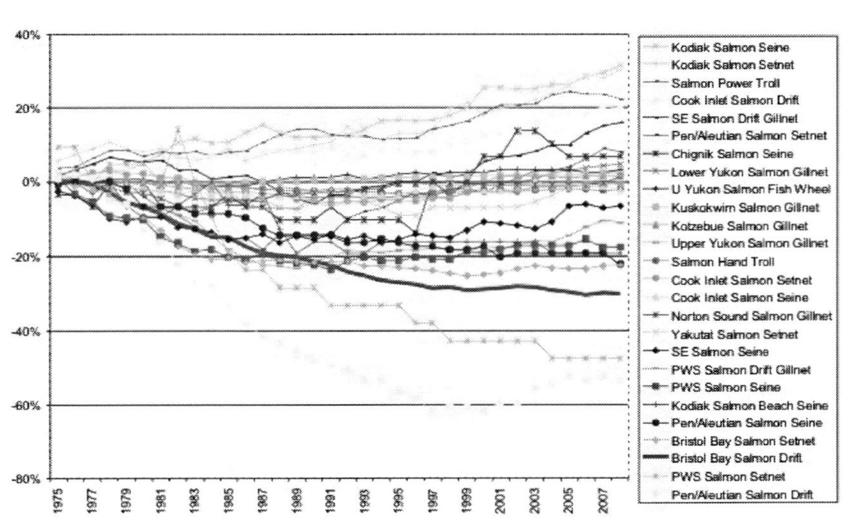

There has been a significant decline in the number and share of permits held by local Bristol Bay residents:

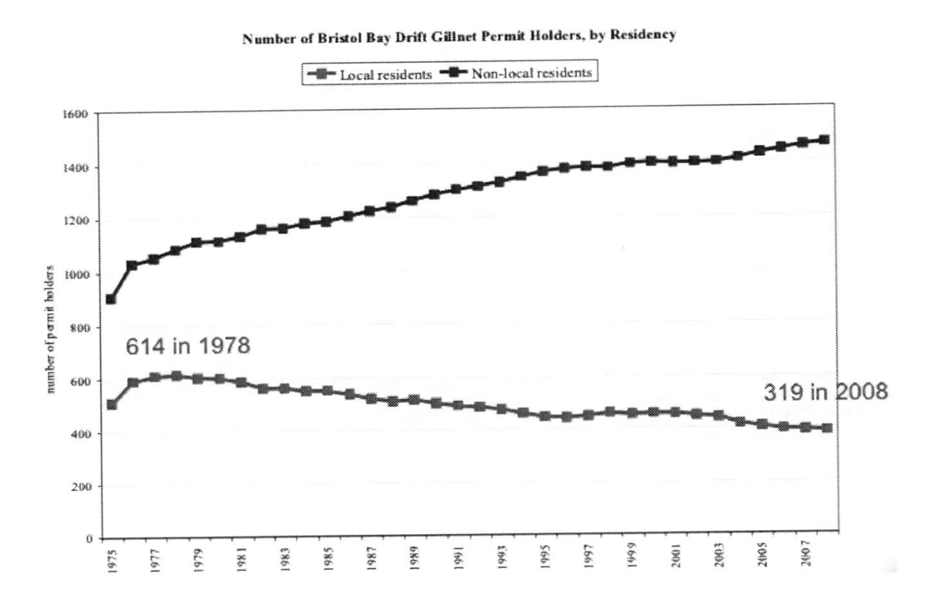

A study conducted in 1984 found that Native residents of Bristol Bay accounted for a disproportionately large share of early permit losses (see table). I don't know of any studies since then that have focused specifically on how Native permit ownership has changed since then.

Estimated Changes in Native and Non-Native Permit Ownership in Bristol Bay Salmon Fisheries Between Initial Issuance and 1983

	Fishery	Native	Non-Native	Total
Initial permit holders	Bristol Bay drift net	547	95	642
	Bristol Bay set net	405	107	512
Permit holders as of the end of 1983	Bristol Bay drift	465	102	567
	Bristol Bay setnet	303	83	386
Change	Bristol Bay drift	-82	7	-75
	Bristol Bay setnet	-102	-24	-126
Percent Change	Bristol Bay drift	-15%	7%	-12%
	Bristol Bay setnet	-25%	-22%	-25%

Source: Kamali, N. 1984. Alaskan Natives and limited fisheries of Alaska: A study of changes in the distribution of permit ownership amongst Alaskan Natives, 1975-1983. CFEC Report 84-8. Commercial Fisheries Entry Commission, Juneau, Alaska.

Annual changes in permit ownership have not been uniform:

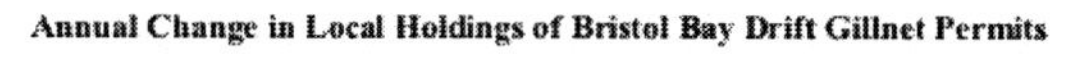

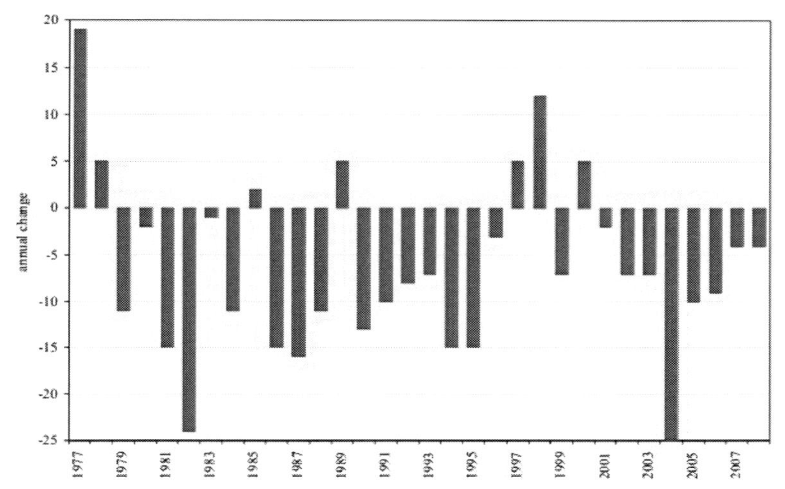

The number of local permits lost from Bristol Bay due to transfers declined after the 1980s.

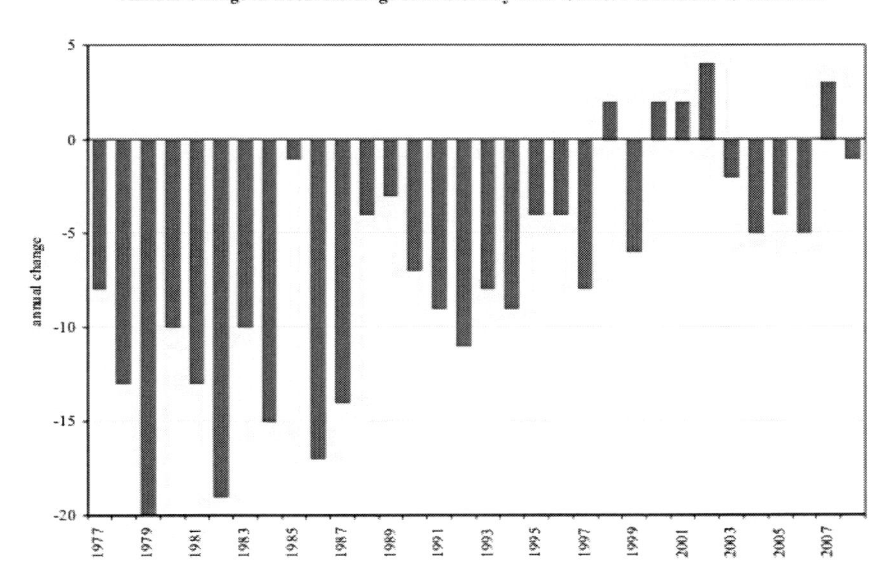

In contrast, number of local permits lost from Bristol Bay due to migration increased after the 1980s.

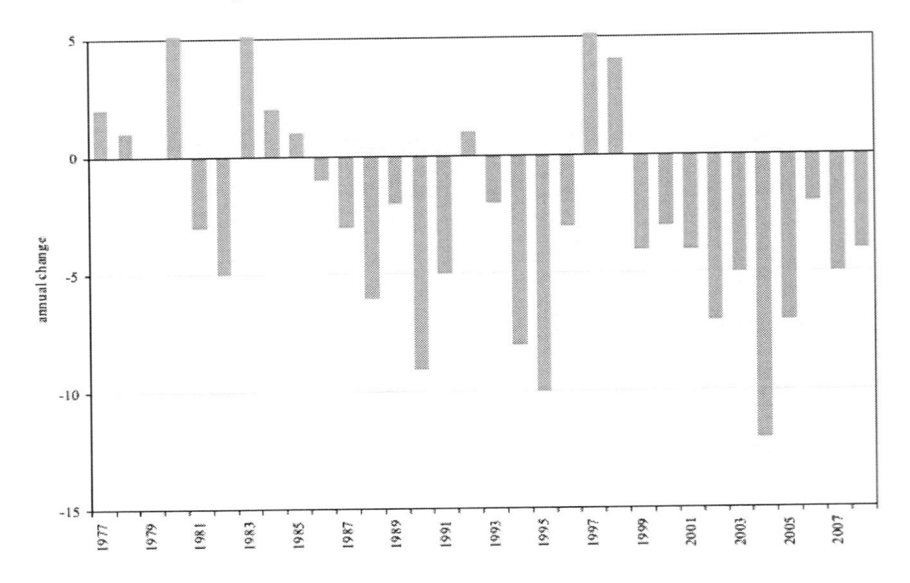

The relative effects of transfers and migration have been changing over time:

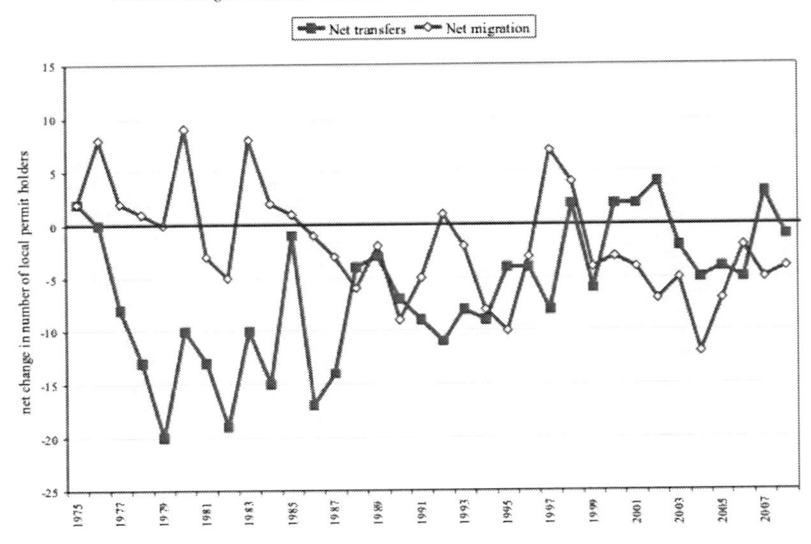

Economic theory of permit transfers

University of Alaska Anchorage anthropologist Dr. Steve Langdon explained the major factors affecting net permit transfers in 1980 (29 years ago!)—just a few years after the limited entry program started: "If a given group of fishermen consistently fall below the average earnings in a fishery, for whatever reason, it is predictable that members of that particular group would be more likely to sell their permits than members of groups who are collectively above the average. "If rural residents have lower average net earnings than other groups, they will be more willing to sell permits than other groups."

Local permit holders have lower average gross earnings than nonlocal permit holders. This helps to explain why they have been relatively more likely to sell permits and relatively less able to buy permits:

In 1980, Dr. Langdon also pointed out that you also have to think about potential permit buyers and whether they can get financing. "Factors on the buying side of importance are the availability of capital for permit purchases, and the ability of rural residents to meet requirements necessary to gain access to financing for permit purchases. Rising cost of technology and permits both will make outright purchases of permits less and less possible so that entry into the fisheries will become more and more dominated by the availability of financing. . . . Financing requirements of the private sector, as well as the . . . operation of the State loan program to date revealed a gloomy picture indeed of access to financing by rural residents."

Part of what matters in the balance of transfers between locals and nonlocals:

- The relative prices that local and nonlocal potential sellers (permit holders) would be willing to sell their permits for.
- The relative prices that local and nonlocal potential buyers would be willing and able to pay for permits.

The relative prices that local and nonlocal potential sellers (permit holders) are willing to sell for are affected by many factors:

- How much they can earn from fishing, which is in turn affected by:
    - ◇ How skilled they are at fishing.
    - ◇ How aggressively they fish.
    - ◇ What kind of boat they have.
    - ◇ What kind of boat investments they can afford.
    - ◇ What it costs them to go fishing—including the cost of getting themselves and their crew to Bristol Bay.

- How much they could earn doing something else if they weren't fishing (their "opportunity cost").
- How much they need cash.
- Non-economic factors such as how much they enjoy fishing.
- The extent to which they prefer to transfer permits by gift or inheritance rather than selling.

Similarly, the relative prices that local and nonlocal potential buyers are willing and able to pay for permits are affected by many factors:

- How much they could earn from fishing.
- How much they could earn doing something else.
- Non-economic factors such as how much they enjoy fishing, and
    - ◇ Their access to capital to pay for permits and boats!
    - ◇ How much money they have that they could invest.
    - ◇ How much they can borrow.
    - ◇ The rate at which they can borrow.

Something else that matters in the balance of transfers between locals and nonlocals is the relative numbers of local and nonlocal potential sellers (permit holders) and potential buyers (non-permit holders.) For example, if there are 1,000 nonlocal potential buyers and only 10 local potential buyers, it is more likely that a nonlocal buyer will offer the highest price than if there are 500 of each. More generally, if the local population is small compared to the number of nonlocals interested in participating in a fishery, it is more likely that transfers will be to nonlocals.

Over time, as the number of locally held permits declines, we would expect net permit transfers to decline because there are fewer potential local sellers. As the number of potential local sellers declines, it is more in balance with the number of potential local buyers. This may be part of the reason why net transfers have declined over time.

Over time, changes keep happening that affect all of these different factors affecting permit transfers:

- The relative numbers of local and nonlocal permit holders change.
- The relative numbers of local and nonlocal potential buyers change.
- The current permit holders get more experience.
- The current permit holders get older (and some die).

- Catches—and catches people expect in the future—go up and down.
- Prices—and prices people expect in the future—go up and down.
- Costs—and cost people expect in the future—go up and down.
- Opportunities in other fisheries get better or worse.

All these changes may affect local and nonlocal potential sellers differently, and all these changes may affect local and nonlocal potential buyers differently.

Economic theory of permit migration

Permit holders may move without transferring for many reasons. Some reasons may have nothing to do with the fishery or whether they are permit holders. They may move because of comparative local and nonlocal school quality, or local vs. nonlocal cost of living. Successful permit holders may in some cases be more likely to migrate than other local residents, for example if they can still come back and work in the region, or if they can earn enough to live elsewhere. But successful permit holders may in some cases be less likely to migrate than other local residents because they can earn a good living while living in the region.

Policy implications

Changes that make the fishery more profitable or favor people who are able to invest more in the fishery tend to increase the probability that permits will be sold from locals to nonlocals.

As the fishery becomes more profitable:

- Travel costs are less of a disadvantage for nonlocals.
- Other fishing and work opportunities for nonlocals are less attractive.
- Permits and boats become more expensive, favoring nonlocals who can borrow money more easily.
- Nonlocals are more likely to be able to invest more in the fishery because:
    - They have more assets they can use as collateral.
    - They are more likely to have friends or relatives who can loan them money.
    - It is easier for them to do business with banks.

The following paradox is disturbing: We want fisheries to be profitable. We want fisheries to be fished by local residents (and Alaskans). But the more profitable a fishery is, the more likely permits are to be transferred from local to nonlocal residents (and non-Alaskans). It is not obvious how to achieve both of these objectives.

Alaska's Next Generation of Fishermen: Understanding Factors Affecting Local Fisheries Access in Coastal Alaska

Rachel Donkersloot
Alaska Marine Conservation Council, Anchorage, Alaska

Courtney Carothers
University of Alaska Fairbanks, Anchorage, Alaska

In 2014, Alaska's newly elected Governor Walker assembled a Fisheries Transition Committee to assist his administration with addressing the major challenges affecting Alaska fisheries. The committee identified "prioritizing and improving fishery access for Alaskans" as one of its key goals. This was the most recent call to arms in a state that provides more than 55% of US seafood production but continues to suffer from the loss of local fisheries access (ASMI 2013).

Declining access to commercial fisheries is not a new problem in Alaska but it is an increasingly pressing one (BBEDC 2007, Knapp 2011, State of Alaska 2012). Coastal communities, fishery managers, and researchers have been grappling with the question of how to ensure the sustained participation of rural coastal residents in Alaska fisheries since the state began limiting entry into commercial fisheries more than 40 years ago (see Langdon 1980, Koslow 1986).[1] Between 1975 and 2014, Alaska's rural fishing communities felt the net loss of more than 2,300 (limited entry) fishing permits (CFEC 2015a:11). Compare this to shifts in other categories of permit holders including urban nonlocals

This article is published with permission from *Environment Magazine*.

1. In 1973, the Alaska State Legislature enacted Alaska's Limited Entry Act (AS 16.43) creating the state's limited entry program. Today, a total of 65 commercial fisheries, including 26 salmon fisheries, are managed under the limited entry system (CFEC 2015a).

and nonresidents of Alaska, which respectively saw a net increase of 309 and 222 permit holders within the same period (ibid).[2]

A similar scenario is unfolding just beyond the three-mile boundary line that marks the outer limit of state waters. Federal fisheries prosecuted off of Alaska's coast are increasingly managed under some form of a catch share (or individual fishing quota [IFQ]) program. Catch share programs typically transform the right to fish into a tradable commodity linked to a specific portion of the total allowable catch (TAC). These programs, which include Bering Sea crab fisheries and the halibut and sablefish fisheries, have resulted in a number of negative consequences for Alaska's rural fishing communities, including consolidation of fishing fleets, concentration of wealth, absentee ownership and leasing of quota, crew job loss and lower crew pay, and rural-to-urban migration of fishing rights. For example, since the implementation of the halibut and sablefish IFQ program in 1995 the number of fishermen in small, rural Gulf of Alaska fishing communities holding quota in these fisheries has declined by 50% (NOAA Fisheries Service 2015:47).

Out with the old, in with outsiders? The graying of the fleet in Alaska fisheries

The exodus of fishing rights from Alaska's fishing towns and villages is compounded by persistent aging and succession trends, a problem referred to as the "graying of the fleet." In 2014 the average age of a state fishing permit holder was 50 years, up from 40 years in 1980 (CFEC 2015a:36). At the same time, the number of Alaska residents under the age of 40 holding fishing permits has fallen from 38% of the total number of permits in 1980 to 17% in 2013. Troubling aging trends are especially pronounced in rural fishing communities. For example, over the past 40 years there has been an 87% decline of young people under the age of 40 who hold salmon seine permits in the rural villages of the Kodiak archipelago. In the Bristol Bay region, the average age of local Bristol Bay salmon drift permit holders has risen from 42.7 to 50.6 years between 1975 and 2013; however, given that nonlocals make up the majority of this fleet, the overall demographic shift among the fleet has been comparatively minor, increasing by only two years, from 45.5 to 47.6 years. The average age of nonresident permit holders in the fishery has actually decreased slightly from 48.3 to 46.9 years.

As the bulk of Alaska permit holders approach retirement age, the big question becomes how the succession of access rights will further exacerbate the exodus of valuable fishing privileges from Alaska's rural

2. The Commercial Fisheries Entry Commission classifies permits based on where the permit holder resides. Permit holders who reside in Alaska are classified into "rural" or "urban" and "local" or "nonlocal" groups. Nonresidents of Alaska are classified into a single category. Categories include Alaska Rural Local, Alaska Rural Nonlocal, Alaska Urban Local, Alaska Urban Nonlocal, and Nonresident (CFEC 2015a:9).

fishing communities. There is growing concern that the majority of these rights will not wind up in the hands of local, and especially young, residents of Alaska's rural fishing communities. What does the loss of fisheries access, income, and identity mean for the long-term sustainability of Alaska's rural fishing communities, cultures, and economies?

Defining the problem in Bristol Bay and Kodiak archipelago fishing communities

To address these questions we launched a research project to examine the problem of the graying of the fleet in two of Alaska's most vital fishing regions: Bristol Bay and the Kodiak archipelago. Although both regions are recognized as primary centers of commercial fishing activity within the state, Bristol Bay and Kodiak Island are supported by economically and ecologically diverse marine ecosystems. The Kodiak Island region boasts a diversified, year-round commercial fishing industry. Over one third of all jobs in Kodiak are directly connected to the fishing industry. Over 27 major fisheries are pursued, including crab, halibut, herring, groundfish, and salmon, and include a diverse fleet of large and small vessels representing multiple gear groups (e.g., trawl, pots, longline, jig, dive) (Kodiak Chamber of Commerce 2013). Bristol Bay fishing communities, on the other hand, are predominantly single-resource dependent, relying largely on a feverish seasonal economy set into motion each summer by world renowned sockeye salmon runs that return to the region's rivers and streams.

Although there are notable demographic and socioeconomic differences between Bristol Bay and Kodiak Island fishing communities and economies, the economic hub and village communities within these regions share similar challenges and vulnerabilities stemming from lack of sustained community participation in local fisheries. Both regions have suffered dramatic declines in permit holders under the age of 40 since 1980 (–47% for Bristol Bay; –48% for Kodiak archipelago) (CFEC 2014). More broadly, recent studies by Carothers (2010, 2015) note that participation in commercial fishing in rural Kodiak archipelago communities has decreased by approximately 70% between 1970 and 2000. In the Bristol Bay salmon fishery, local permit ownership declined by 50% between 1975 and 2014 (from 1,372 to 686 local permit holders) (CFEC 2015b). Clearly, these regions—recognized and celebrated for high levels of fisheries dependence and engagement—are compelling places to explore how and why local fishing livelihoods and communities are impacted by problems underlying and arising from the graying of the fleet.

Our project aims to document and compare perceived and experienced barriers to entry into, and upward mobility within, fisheries among local youth and young fishery participants in fishing communities in the Bristol Bay and Kodiak archipelago regions. As part of this effort, we examine geographic, economic, social, and cultural factors influencing young people's attitudes toward and level of participation (actual and desired) in Alaska fisheries. We draw on ethnographic, interview, and survey data to address these objectives. To date, our project team, which includes co-principal investigator Paula Cullenberg, director of Alaska Sea Grant, and two University of Alaska graduate students (Jesse Coleman and Danielle Ringer), has completed 130 interviews with local and young fishermen in our study communities.[3] More than 800 middle and high school students attending Bristol Bay and Kodiak Island schools completed a survey administered in spring 2015.[4] In this article we discuss key themes and project findings that have emerged from initial data analyses. A more exhaustive analysis of project data is currently under way, the results of which will be shared on our project website: http://fishermen.alaska.edu.

Beyond the bank: barriers to entry and upward mobility in Alaska fisheries

The out-migration of access rights from rural Alaska fishing communities is frequently conceptualized as a primarily economic problem stemming from rural residents' limited access to financing, capital, and credit when compared to their urban and non-Alaska resident counterparts (see Knapp 2011). Undoubtedly, rising entry costs are one of the most formidable effects of limited access programs such as catch share programs and Alaska's Limited Entry Program. Rosvold (2007) estimates that fishing access rights (permits and quota) now make up about 83% of the value of a typical fishing operation in Alaska, which for a vessel that fishes for salmon, crab, sablefish, and halibut he estimates to be over $4.5 million. The high cost of fishing rights is certainly working to impede upward mobility in fishing careers for crew members and skippers and create seemingly impassable barriers to entry for new and young fishermen (Olson 2011, Carothers 2015).

As expected, the high cost of fishing rights emerged as a major theme from interviews with new, young, and local fishermen in our study communities. In Bristol Bay, for example, the 2014 estimated value of a salmon drift permit was about $150,000 (a setnet permit is valued around $40,000). In Kodiak, the cost of a salmon seine permit averaged $50,000. Halibut IFQ in the central Gulf of Alaska (Area 3A) was selling

3. Kodiak Island communities include Kodiak City, Old Harbor, and Ouzinkie. Bristol Bay study communities include Dillingham, Togiak, Kokhanok, and the Bristol Bay Borough (Naknek, South Naknek, and King Salmon).

4. The survey focused on young people's perceptions and experiences of the local fishing industry, life in their home community, and plans for the future.

at over $50 per pound at the end of 2014.[5] These costs, paired with other economic and ecological drivers of change such as increasing vessel and gear costs, poor fish prices, and in some cases low stock abundance can thwart the ambitions of the next generation of commercial fishermen.

> Well, in all reality, I still would like to be drifting, but I didn't have a [drift] permit of my own, I just always captained someone else's boat... I mean, if you want a new boat or even a decent boat, it's a couple hundred thousand dollars, and then right now, you're looking at another probably couple hundred thousand in gear... I couldn't afford to buy a boat and permit and stay involved in the fishery at an owner level, so I opted to set netting. (*Permit holder, salmon setnet, Bristol Bay region, February 2015*)
>
> We seriously looked at [buying into the fishery]... [But] everything out here, the price of everything just went up, and the price of salmon was going down. I mean it was your maintenance costs and everything else. That was just getting horrendous. You know, we were discussing different options and different things [to buy in] and we both have made that decision that there's a lot to it, and it's just not really happening right now... It would be just the initial cost. *(Retired crew, salmon drift, Bristol Bay region, May 2015)*
>
> The costs are just astronomical when you're really young. Trying to secure loans that are $500,000 or more is just—nobody's just going to hand it over to an 18 year old. *(Permit holder, salmon drift, Bristol Bay region, October 2014)*
>
> [You] better have understanding parents or a really friendly uncle willing to loan you enough money to do it. There's no really realistic way for anybody of any age that you would even consider young, to own enough collateral for a bank to consider giving them such a high risk loan... At one point in time it was pretty much anybody that [had] a skiff and wanted to go fishing could. And now regulations changed so much that there's not really any point. Unless you just happen to have an extra half a million dollars kicking around. There's no entry-level... there's no way to make a living at it anymore. Ah... you can make a living, but you're never gonna get ahead in life. *(Crew, pot and longline fisheries, Kodiak region, February 2015)*

While the escalating cost of entering a fishery and operating a competitive fishing business in coastal Alaska was a pervasive theme in interviews, other factors were also linked to declining local fisheries participation. Many project participants referenced a lack of fishing experience and knowledge among young residents as an important component of the problem. In the excerpts below, participants reflect on the importance and diminishing nature of learning opportunities for young people to gain necessary fishing skills and knowledge in rural communities.

5. CFEC 2015. https://www.cfec.state.ak.us/pmtvalue/mnusalm.htm. Accessed December 21, 2015.

I think a lot of youth would like to go fishing, they'd like to be a part of it, but I see there being less opportunity, just because your permit holders aren't there in the village, a lot of people have sold out... A lot of kids in the village here, they're not experienced... You get to be, you know, 16, 18 years old and you have no experience. *(Permit holder, salmon setnet, Bristol Bay, February 2015)*

Because they don't have the experience... They just don't have the knowledge—maybe because their parents didn't teach them. They just said, "Here, have my permit." They didn't teach them how to do stuff. You know—like work on their motor, something's clogging the gas or something, you know... *(Permit holder, salmon setnet, Bristol Bay, January 2015)*

I think I was really lucky that I had who I consider my mentor, approach me and say, 'Hey, if this is something you want to do, I will help you. I will show what to do.' Had I not had that, I probably still would be a deckhand... I had a permit but I didn't have the knowledge, I didn't have the equipment... So I think that was key. The biggest reason that I am fishing today was just because I had somebody to show me the ropes. I think that's one of the biggest things that younger people coming into the fishery have to face. *(Permit holder, salmon setnet, Bristol Bay, March 2015)*

You don't see the younger kids as much anymore... [In] high school, even starting in junior high, we—almost all my buddies—we had fishing jobs. And now as a captain, besides some of the owners that have their sons, you don't see a lot of 14, 15, 16-year-old crew members anymore. And I think that some of that is the school system and some of that is lack of desire from the kids, too... [Back then] all of us were driving new pickups and had our spending money but it was because like I said, [in] most of the families, somebody in the families was fishing. It was pretty commonplace. *(Permit holder, multiple fisheries, Kodiak region, September 2014)*

Our survey data also give support for this trend of decreasing youth engagement in commercial fishing, especially in the Kodiak region. Only 9.4% of high school students surveyed in the Kodiak area have ever been involved in the commercial fishing industry (although 33% have family ties to commercial fishing).[6] In the Bristol Bay area, 48% percent of high school students have been engaged in commercial fishing (although 72% have family ties to commercial fishing).

Others participants cited a lack of alternative fishery and non-fishery employment opportunities in their communities as a factor affecting local fisheries participation. In Kodiak, where so many fisheries are pursued, the commodification of access rights has made it particularly difficult for young or new fishermen to diversify into multiple fisheries when they are starting out. Very few fisheries remain open to newcom-

6. The off-road system Kodiak villages in our study show more fisheries engagement (21% of youth surveyed in Ouzinkie have had some commercial fishing engagement; 91% of youth in this community have family ties to fishing); 56% of youth surveyed in Old Harbor have commercial fishing engagement (89% report family ties to fishing).

ers to move into owner-operator roles without substantial investment. The cod jig fishery is one example of a fishery that is relatively easy to gain access to; however, as study participants noted, it is very difficult to operate a profitable business from jigging alone. Outside of fishing, local employment opportunities were deemed essential to support and supplement seasonal fishing livelihoods. This was especially the case in Bristol Bay due to the short (roughly six weeks) salmon season. At the same time, having the ability to take time off from work to prepare for and participate in local fisheries was identified as a growing constraint.

> The toughest part right now is—like for me, for my job, I can't go fishing. It's so busy in the summer that I don't have time off and this job pays, and it pays well, and it lets me do the kinds of things I like to do anyway. And I get, you know, year-round employment. You can't really put a price on that with the benefits, insurance and everything. *(Retired crew, salmon drift, Bristol Bay region, May 2015)*
>
> I have to make sure I take enough time off from my other job. It's mostly just time management, having the right gear and having crew to take care of the fish, if and when you get them. *(Permit holder, salmon setnet, Bristol Bay region, February 2015)*
>
> Well, you hear some companies don't want to hire local people if they're fishermen... Because they don't want them to be leaving their job in the summertime. And so you hear stories about—I guess when they interview they ask you, 'Do you commercial fish?' *(Permit holder, salmon setnet, Bristol Bay region, May 2015)*

Rising social problems, especially related to drug and alcohol abuse, also emerged as a barrier that can disrupt and deter pathways to productive fishing careers and permit ownership. Participants described "rough family situations," costly legal troubles, and other social ills and repercussions partially though not entirely stemming from alcohol and the more recent arrival of hard drugs, especially meth and heroin. In response to an open-ended survey question, over 56% of middle and high schools students surveyed in Bristol Bay communities stated that drugs and alcohol are the biggest concerns they have about their communities. About 38% of students surveyed in Kodiak share this concern. Such instances were described as impediments to establishing ownership-level fishing careers as well as social scourges detrimental to the overall health and well-being of the community and fisheries at large.

> When I was little, I remember it being completely different... I think it's a pretty rough place right now. And so I don't think I would want to live here right now. I think there's better places. *(Crew, salmon drift, Bristol Bay region, July 2014)*
>
> Whether it's drugs or alcohol, I've seen some of that out there [in the community], where you're struggling in life also and it kind of translates into your career. So if you're an alcoholic that can't function, then it's hard to keep crew, it's hard to keep fishing, it's ... definitely a struggle. *(Crew, salmon seine, Kodiak region, September 2014)*

> I was supposed to get a boat this year, and a motor. But I ended up getting caught up in the legal system, so that didn't happen for me. I had to use what [money] I had. *(Permit holder, salmon drift, Bristol Bay region, April 2014)*
>
> I'd rather live there up in the village than here [in Anchorage], because it's just so—you're not free [in the city]. You can't go out and snow machine out your door or go shoot something or go catch something without a permit. So I'd rather live [in the village]. But we're living here because of the jobs during the wintertime... I do want to go back, but it's getting so bad there... [The] drugs—drugs are really bad. *(Permit holder, salmon setnet, Bristol Bay, January 2015)*

Taken in sum, the above examples begin to show how local fisheries access is underpinned by myriad, complex and intersecting factors. Access to capital is clearly an important dimension of the problem of the graying of the fleet but economic explanations alone provide at best a partial understanding of the problem. Interviews with young local fishermen help to widen the aperture on our understanding of the graying of the fleet by shedding light on the ways in which young people's attitudes toward, and pathways to, permit ownership are supported and constrained by place-based dynamics, including social and cultural resources and barriers. More broadly, what these accounts help to capture is how young people's sense of place and attitudes toward one's home community informs one's decision to stay or leave (Donkersloot 2007, 2010, 2011). For example, only 26% of students in the Kodiak archipelago, and 22% of students in Bristol Bay agree that the future looks good for young people who stay in their home communities. Understanding youths' desires for remaining in or leaving their communities is important because a growing number of permits are leaving rural Alaska through migration processes versus permit sales.

Migration of permits, migration of people

Recent data show that migration (i.e., the relocation of permit holders) has changed the resident/nonresident balance in Alaska fisheries to a greater degree than permit transfers/sales (CFEC 2015). At the end of 2014, the net result of permit transfer activity had decreased the number of permits held by nonresidents by 344 permits, while permit holders moving into and out of Alaska resulted in a net increase of 1,056 nonresident permits (CFEC 2015a:7). Looking specifically at rural local permits one sees a net loss of 1,202 through migration. Compare this to a net loss of 224 rural local permits through permit transfer activity (CFEC 2015a:11).[7] How fishing rights leave a region or community varies from place to place with some communities more susceptible to vulnerabilities associated with population decline than others.

7. Another 878 rural local permits have been cancelled (CFEC 2015a:11).

For example, all but one of our study communities experienced population decline between 2000 and 2010.[8] In some cases, such as Kodiak City (–3.2%), Dillingham (–5.6%), and the villages of Kokhanok (–2.3%) and Old Harbor (–8%), this loss was slight to moderate. Other communities suffered dramatic declines in local population. These include Naknek (–19.8%), South Naknek (–42.3%), and King Salmon (–15.4%) in Bristol Bay and the village of Ouzinkie (–28.4%) in the Kodiak archipelago. Population shifts are driven by many factors and motivations, including employment and career opportunities, retirement, family choices about schools and education, climate, cost of living, and quality of life. What is clear is that factors affecting population change in rural coastal Alaska also affect local fisheries access. For example, in the last ten years (2005-2014) the Bristol Bay Borough lost 26 local salmon (drift and setnet) permits through out-migration with the bulk (n = 19) of these coming out of the small village of South Naknek. This is significant; it represents the loss of 26 small businesses.

More than money: understanding the importance of local fishing livelihoods in rural Alaska

The economic impact and loss of local income stemming from declining access to local fishery resources is reason enough to be alarmed at recent trends. But loss of local fisheries access and participation is disconcerting for many reasons.

Several studies describe the dynamic relationship between commercial fisheries participation and subsistence practices in the mixed economies of rural Alaska (Langdon 1991, Wolfe et al. 2009). Drawing on the seminal work of Robert Wolfe and colleagues, Davin Holen (2014:10) writes that rural "households with fishing permits are often also the households that are high producers of subsistence foods. A household's wild food harvest increases by 125.8% if the household is also involved in commercial fishing." Holen (ibid.) continues: "The harvest of wild foods in rural Alaska remains a key factor for providing for food security, but the subsistence economy is intimately tied to the cash economy, leaving rural communities in Alaska vulnerable, especially with a declining participation in commercial fishing by rural residents." In their work in southern Bering Sea villages, Reedy and Maschner (2014:13) further demonstrate the importance of local commercial fisheries participation to subsistence harvests and local food sharing networks, primarily because "fishermen who participate in commercial fisheries are often the most important providers in all sharing networks and are high status members of their communities." The social, cultural and economic value of subsistence foods and harvesting

8. The population of Togiak grew by 1% between 2000 and 2010 (US Census Bureau 2012).

practices emerged as one of the most salient themes in the rural villages included in this study. In many interviews, the natural environment and ability to harvest one's own food was what young people valued most about life in their home communities. But commercial fishing in and of itself was also identified as integral to the reproduction of local social and family relationships, identities and cultural values.

> It's not always about the money. It's enjoying family time—that's what we were doing. A lot of family time. And working together and just depending on each other during that time. *(Permit holder, salmon setnet, Bristol Bay region, April 2015)*
>
> It gives you your identity. For me it's just part of my home and where I am and where I live and it gave me [a] really good strong work ethic. And I think that's one of the things that a lot of kids are lacking these days... I value all the years that I had on the boat and the time that I had with my family... It was a family operation and you just can't take that time ever back—it's pretty amazing. *(Retired crew, salmon seine, Kodiak region, May 2015)*
>
> I remember my grandpa used to say many years [ago]... the fishing will still be [here] as long as there's generation to keep it. [It's] hard to translate... I'm kind of losing my touch at speaking Yupik [but it's about keeping fishing] alive, and staying in the community and thriving... If not for the economy, for the village. If not for the village, for the family. *(Permit holder, salmon setnet, Bristol Bay region, January 2015)*
>
> I think regardless of whether or not it's a big money maker, it's part of who we are and the opportunity for people who haven't been involved or younger people who are just getting involved, I would like to see opportunity for that. Just being involved in that cycle... You're part of something a lot bigger—you know, you do go down and work hard, you may not make all of your money in the year, but it's self-sufficiency and reliance and I think it gives people a better sense of themselves. *(Permit holder, salmon setnet, Bristol Bay region, February 2015)*

Concluding remarks: sustaining places

There is no single silver bullet or immediate solution that can address the ongoing flight of fishing rights from Alaska's rural fishing communities or the lack of access opportunities for young fishermen. State and community leaders are acutely aware of this. We have our work cut out for us when it comes to tackling one of the major challenges affecting Alaska fisheries. The lack of Alaskans and especially rural young Alaskans participating in local fisheries threatens the long-term sustainability of our fishing communities. We find ourselves at an increasingly critical moment given the rising age of the fleet and succession trends that will trigger greater loss of access for rural Alaskans. Identifying potential policy responses that may help to reverse these trends is also a key objective of our study. We are in the final phase of a global review of programs and regulatory frameworks that can help inform efforts to improve fishery access for Alaskans. Other countries have implemented youth permits; special provisions for indigenous, rural and small-scale fisheries; recruitment quotas; and other programs that have been designed to address these issues of access that often emerge when the rights to fisheries are monetized. The problem is complex but not insurmountable. Solutions may be costly in terms of time, capital, and political will, but not nearly as costly as the price of inaction.

References

ASMI (Alaska Seafood Marketing Institute) 2013. Economic value of the Alaska seafood industry. Report prepared by McDowell Group. http://pressroom.alaskaseafood.org/wp-content/uploads/2013/08/AK-Seafood-Impact-Report.pdf

Bristol Bay Economic Development Corporation (BBEDC) Permit Loan Committee. 2007. Restoring salmon permits to Bristol Bay residents. Report prepared by RedPoint Associates and Alaska Growth Capital for BBEDC.

Carothers, C. 2010. Tragedy of commodification: Transitions in Alutiiq fishing communities in Alaska. Maritime Studies (MAST) 2010: 90(2):91-115.

Carothers, C. 2015. Fisheries privatization, social transitions, and well-being in Kodiak, Alaska. Marine Policy 61:313-322. http://dx.doi.org/10/1016/j.marpol.2014.11019

CFEC. 2014. Analysis of data provided by Alaska Commercial Fishing Entry Commission conducted by authors. Not available.

CFEC. 2015a. Executive summary: Changes in the distribution of Alaska's Commercial Fisheries Entry Permits, 1975-2014. Alaska Commercial Fishing Entry Commission (CFEC) Report 15-03N EXEC.

CFEC. 2015b. CFEC permit holdings and estimates of gross earnings in the Bristol Bay commercial salmon fisheries, 1975-2014. Alaska Commercial Fishing Entry Commission (CFEC) Report 15-4N.

Donkersloot, R. 2007. Youth emigration and reasons to stay: Linking demographic and ecological change in Bristol Bay, Alaska. In: P. Cullenberg (ed.), Alaska's fishing communities: Harvesting the future. Alaska Sea Grant, University of Alaska Fairbanks, pp. 73-79.

Donkersloot, R. 2010. The politics of place and identity in an Irish fishing locale. Journal of Maritime Studies 9(2):33-53.

Donkersloot, R. 2011. What keeps me here": Gendered and generational perspectives on home and migration in a rural Irish fishing locale. PhD dissertation, Department of Anthropology, University of British Columbia, Vancouver, BC.

Holen, D. 2014. Fishing for community and culture: The value of fisheries in rural Alaska. Polar Record 50(4):403-413. http://dx.doi.org/10.1017/S0032247414000205

Knapp, G. 2011. Local permit ownership in Alaska salmon fisheries. Marine Policy 35:658-666. http://dx.doi.org/10.1016/j.marpol.2011.02.006

Kodiak Chamber of Commerce. 2013. Kodiak community profile and economic indicators report. 4th Quarter. Kodiak Chamber of Commerce, Kodiak, Alaska.

Koslow, A. 1986. Limited entry policy and impacts on Bristol Bay fishermen. In: S. Langdon (ed.), Contemporary Alaskan Native economies. University Press of America, Lanham, MD, pp. 47-62.

Langdon, S. 1980. Transfer patterns in Alaskan limited entry fisheries. Final Report for the Limited Entry Study Group of the Alaska State Legislature.

Langdon, S. 1991. The integration of cash and subsistence in southwest Alaskan Yup'ik Eskimo communities. Senri Ethnological Studies 30:269-291.

NOAA Fisheries Service. 2015. Report on holdings of individual fishing quota (IFQ) by residents of selected Gulf of Alaska fishing communities 1995-2014. November 2015. https://alaskafisheries.noaa.gov/sites/default/files/reports/ifq_community_holdings_95-14.pdf. Accessed December 30, 2015.

Olson, J. 2011. Understanding and contextualizing social impacts from the privatization of fisheries: An overview. Ocean and Coastal Management 54(5):353-363. http://dx.doi.org/10.1016/j.ocecoaman.2011.02.002

Reedy, K., and H. Maschner. 2014. Traditional foods and corporate controls: Networks of household access to key marine species in southern Bering Sea villages. Polar Record 50(4):364-378. http://dx.doi.org/10.1017/S0032247414000084

Rosvold, E. 2007. Graying of the fleet: Community impacts of asset transfers. In: P. Cullenberg (ed.), Alaska's fishing communities: Harvesting the future. Alaska Sea Grant, University of Alaska Fairbanks, pp. 67-72.

State of Alaska. 2012. HCR18—Commercial Fisheries Programs. http://www.legis.state.ak.us/PDF/27/Bills/HCR018C.PDF. Accessed June 2016.

US Census Bureau. 2012. Alaska: 2010 census of population and housing unit counts. CPH 2-3. https://www.census.gov/prod/cen2010/cph-2-3.pdf

Wolfe, R.J., C.L. Scott, W.E. Simeone, C.J. Utermohle, and M.C. Pete. 2009. The "Super-Household" in Alaska Native subsistence economies. Final report to the National Science Foundation, Project ARC 0352611.

Defining the Policy Problem: A Community-Based Fisherman's Perspective

Jeff Farvour
Sitka, Alaska

Introduction

I moved to Sitka from Tacoma, Washington, in 1995 but I've been fishing in Alaska since 1989. Depending on which definition you look at, Sitka is rural. It is remote with no roads in or out, fulfilling the federal definition for subsistence. I moved to Alaska because I believed that if I was going to continue to make a living commercial fishing in Alaska's waters, I needed to give something back. That's why I moved to Sitka—to pursue a lifestyle of commercial fishing living year-round in a coastal Alaska fishing community.

I do not come from a fishing family, nor did I receive an initial fishing allocation or permits. I found good safe boats to fish on, and saved my crew shares to buy small blocks of halibut IFQ (individual fishing quota), a limited entry troll permit, and a troller. I also did some shipwrighting in the winters, but truth be told, commercial fishing often subsidized my shipwright habit!

I have a 40 foot salmon troller, and I fish halibut both from a skiff and my troller. I have crewed for the last 12 years on a productive 46 foot longline/troller. I crewed before that on other halibut and blackcod boats, factory longliners in the Bering Sea, the Bristol Bay driftnet fishery, Southeast salmon seine, and others.

Policy obstacles to access

Agri-biz model

Under the agri-biz model there is lots of pressure to get big or go home. This puts unrealistic expectations on the fishermen in the fishery and on new entrants, that you have to be big, have multiple boats, and be highly consolidated to be successful. Overcapitalizing can become a reason for policies that favor larger operations, policies that are not practical for small fishing boats.

For example, Environmental Protection Agency (EPA) discharge permits that are designed for very large vessels, floating factories, and shipping tankers are not practical for small vessels. This would require every commercial vessel regardless of size to have permits for deck hose water, rain water coming off the decks, cleaning up fish guts and slime, fish hold ice melt, and possibly require them to store and bring that water to town for inspection. Senator Lisa Murkowski has been successful at getting temporary exceptions for the fleet but may not be able to do so much longer.

Coast Guard requirements

According to recent NIOSH reports, the Alaska IFQ and Southeast salmon fisheries are among the safest in the nation. Despite that, the Coast Guard has implemented new costly laws that are now in effect for commercial fishing vessels of all sizes. Among other things these include:

- It is illegal to operate your vessel unless you install a survival craft that ensures no part of an individual is immersed in water, on all commercial fishing vessels operating beyond three nautical miles of the baseline.
- Individuals in charge of commercial fishing vessels operating beyond three nautical miles of the baseline are required to keep a record of equipment maintenance, and required instruction and drills.
- Training is required for all individuals in charge of commercial fishing vessels operating beyond three nautical miles of the baseline.
- New vessels, built after January 1, 2010, that are less than 50 feet overall in length are required to be constructed in a manner that provides a level of safety equivalent to the minimum standards established for recreational vessels.
- New vessels, built after July 1, 2012, that are at least 50 feet overall in length and will operate beyond three nautical miles of the baseline must meet survey and classification requirements. Vessels built to class requirements before July 1, 2012, must remain in class.

Structure of limited access programs

Season structure and length can significantly affect Alaska fishing community access to the fishery.

Southeast dive fisheries had a derby, and more and more divers from down south were coming up and sweeping up the quota. To stem the flow, openings were changed from a full-on derby to a one or 1½ day fishery per week. That gave Alaska locals a better chance at competing and made it less attractive to nonresidents, as they would have to spend a longer time in Alaska. Consequently, the price also improved because the market wasn't flooded with all the product at once.

The Southeast summer troll fishery is prosecuted in a way that most of the king salmon quota is taken in a 5-10 day derby in July. The winter fishery is mostly fished by locals and is the time we get the best price for our fish. Some locals have asked for management that would spread out more of the quota in the winter and spring, which would give them better access to fishing and get the best price for the fish. This has been met with adamant opposition by the nonresident component of the troll fleet.

For **IFQ**, an 8-month season made it possible for smaller, local boats to fish more safely.

Owner/operator fleets

Area 2C, known as Southeast Alaska, has the strongest policy for owner on board requirements and as such the harvest is spread out among more vessels and a wider range of vessel sizes.

Owner on board policies promote a strong sense of **stewardship**, commitment to the welfare of the rest of the fleet, strong advocacy of the resource, and fishing community advocacy. They are the ones I see at meetings participating in a wide range of issues (besides the lawyers and lobbyists who work mostly for the corporate operations).

Owner-operated fleets help prevent loss of lifestyle instead of encouraging purely business models.

Leasing

Leasing (and non–owner on board) is a major issue in many of Alaska's fisheries. Leasing and absentee ownership often extract significant amounts of resource rents (ex-vessel) away from the harvesters, leaving fishermen poorly situated to sustain their fishing operations. This competition for fishing opportunity favors the absentee owner, through high lease rates charged by the absentee owner of fishing quota and permits who does not share in the annual risk of fishing. This scenario prevents profits from being reinvested by fishermen and makes it nearly impossible for new entrants to come into vested ownership and live in coastal Alaska fishing communities.

Communities and crew left out of initial allocations

Crew make up the largest portion of the harvesters, yet they are consistently left out of allocations. Crew have bills to pay and families to feed. Not recognizing crew during allocations leaves them without a foundation to invest in fishing.

It is the same with communities. **Communities** are also consistently left out of allocations. Communities need stability to support a livable, enjoyable place to live and raise future fishermen.

Alaska needs independent community-based fishermen invested, and communities invested, to maintain healthy vibrant fishing communities, prevent resources from leaving communities, maintain infrastructure, and ensure future generations have access. Owner-operated fleets are essential to maintain this. It is important to cultivate a healthy dynamic of independent community-based fishermen working with communities.

Consolidation

The halibut/sablefish fishery has experienced about a 50% reduction in the size of its fleet. Consolidation has happened across all vessel sizes, but more in the smaller boats. The cost of halibut quota share has increased 600% in Southeast and southcentral Alaska. This makes it very difficult to enter and continue building a diversified fishing operation.

Ownership caps

Ownership caps, or how much an individual can own, are necessary to control consolidation. In the halibut/sablefish IFQ fishery, we have vessel and use or ownership caps. Area 2C has a lower ownership and vessel cap than the other areas in Alaska and thus a greater percent of the Southeast halibut quota is harvested by a wider range of vessels than in other areas. You can see in the graph, which the Commercial Fishing Entry Commission put together for the Alaska Longline Fishermen's Association, that area 2C has effort spread throughout the fleet. This is compared to 3A in which 50% of the vessels landed about 96% of the 3A halibut. It is a product of 2C having tighter controls on consolidation and an owner on board policy at the time of IFQ implementation.

Policy scaled to work for small boats and communities

Reallocation of halibut quota share from community-based fishermen can occur in two major ways.

1. Fishermen who are invested in the small boat historic halibut fishery need **stability and to reduce risk** in order to be viable. Living with quotas going up and down corresponding with biology is a fact of life, but bycatch of halibut poses a huge risk to halibut fisheries. Since

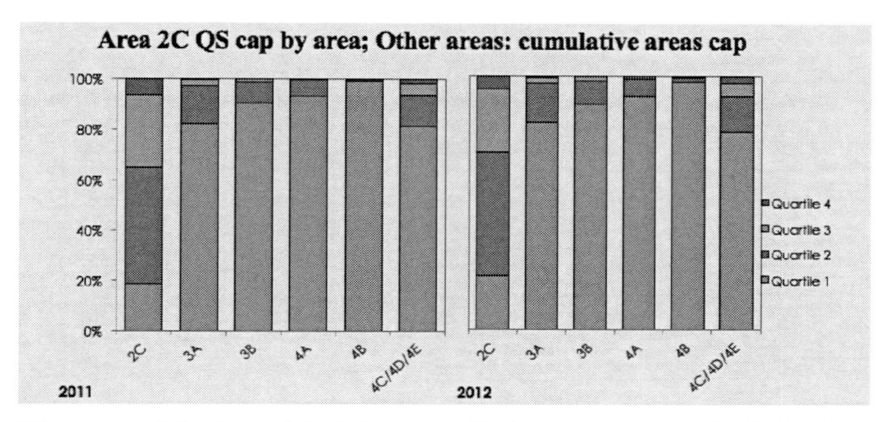

The parts of the bars (labeled as quartiles) each represent 25% of unique halibut fishermen. For example, in 2011 in area 3A, 25% of the fishermen own 80% of the halibut quota. In 2C the quota is more evenly divided up among fishermen.

bycatch of halibut is prioritized over directed halibut fisheries, bycatch of halibut could potentially shut down the halibut fishery. For example, in the Bering Sea bycatch of halibut from the mostly industrial trawl fisheries accounts for 373% more halibut mortality than is allowed as catch by the commercial fishery in that area. Trawl bycatch of halibut in the western and central Gulf of Alaska accounts for more than 3 million pounds of halibut removal in those areas. Trawl bycatch is mostly juvenile fish, which are potential future rebuilders of the stocks, and there is significant migration of Bering Sea halibut to all of Alaska.

2. Although we are fortunate to not have trawling in Southeast we have had many management issues with the charter sector. For example, we had a 20 year battle to settle an allocation with a new and growing **charter** fishery, which could have potentially closed the Southeast halibut fishery and significantly curtailed the 3A halibut fishery. Reallocation is again being considered at the North Pacific Fishery Management Council, but in terms of an RQE (recreational quota entity) that would allow the charter sector to erode the commercial halibut fishery further in 2C and 3A, by allowing the charter sector to buy commercial allocation but not allow the commercial halibut fishery to buy charter allocation. This is especially damaging during these times of low halibut abundance that not only affects the resident halibut fishery, but also has significant impacts on the community infrastructure and support sectors. Most of the 2C halibut quota is held by Alaskans, yet 98% of 2C charter clients are nonresidents.

The Sitka LAMP (Local Area Management Plan) provides local subsistence only in the summer in Sitka Sound, while allowing small commercial halibut vessels fishing opportunity in protected waters in spring and fall.

Multiple fisheries

The very high costs of entry can make participating in multiple fisheries even more difficult. Options become more limited. It is very important to be diversified, especially after limited entry and catch share style policies have been implemented. Being dependent on a single fishery or species can leave one financially vulnerable.

Summary

To summarize, three areas need addressing to improve fishery accessibility: a regulatory framework that works for small boats; limited access programs that promote opportunities for a community-based fleet; and policy scaled to meet the needs of small boat fishermen.

Constitutional Constraints on Community Permit Bank Legislation

Jim Brennan
Brennan and Heideman, Anchorage, Alaska

Community Permit Bank legislation would involve preferences to residents of coastal Alaska communities in leases or similar authorization to use limited entry permits purchased or administered by regional Community Permit Banks. A CPB program would feature preferences over non-Alaska residents **and** over other Alaska residents.

There are constitutional limits on preferences, and against discrimination, under both federal and Alaska constitutions, and they are **different** under each. I'll focus on the proposed CPB program, but these constitutional concerns also apply to other state fisheries programs restricting access and granting preferences.

Some legal analysts say the legislation will easily withstand challenge; other say it will definitely lose. **Both are wrong.** Success against challenge depends on how law is crafted, including stated **purposes** and **legislative findings**. If ultimate legislation remains skillfully crafted, I think it's "odds-on" to survive challenges, which are inevitable.

Part of the title of this workshop is "Charting the Future," and it is fair to say that a CPB program must navigate between two constitutional reefs—one state, one federal. If you limit your attention to one, you may well run hard aground on the other. We have to watch out for several clauses in each constitution.

Alaska constitution

There are three clauses: (1) "common use" of fish resources; (2) "no exclusive right or special privilege" of fishery; and (3) equal protection (nondiscrimination) clauses. Except for challenges to some minor provisions, Alaska's 1973 Limited Entry Act has survived all state constitutional challenges, and its methods of awarding permits have been upheld.

Apokedak upheld initial criteria for issuance, restricting initial entry to existing gear license holders. *Ostrosky* upheld permit sale and transfer provisions.

Of great importance in defending any **new** limited entry law is the 1972 amendment to the Alaska Constitution, **ratified by the voters**, adding a sentence to Article VIII, Sec. 15. After the language saying there can be no "exclusive right" or "special privilege," the amendment added: "This section does not restrict the power of the State to limit entry into any fishery for purposes of resource conservation, to prevent economic distress among fishermen and those dependent upon them for a livelihood and to promote the efficient development of aquaculture in the State."

Ostrosky: This affirmative constitutional authorization of a limited entry system counteracts **all** of the **other** constitutional limitations (e.g., common use, equal protection), too. However, *Ostrosky* and later cases also say there is a "logical argument" that limited entry laws must feature the "least possible infringement" on these other constitutional limitations. But the court has come up short of making this concept an express holding.

Therefore, a CPB statute should seek **protection** of 1972 amendment by showing that it serves at least one **purpose** of the amendment: (1) resource conservation; or (2) **prevention of economic distress among fishermen and those dependent upon them for a livelihood.** Legislative history shows that **economics** was more important than biology in the framers' intent.

Primary task

A primary task is for legislative **findings** to link up with the constitutional **purposes.**

Following are some ideas to carry out that task.

- Find that the fishing economies of coastal Alaska towns depend on new entry by local residents into the permitted fisheries.
- Find that lack of access to permits is a barrier to entry to local residents who could otherwise more economically engage in a fishery than nonresidents, in a way that better benefits local fish industry infrastructure.
- Find that the exodus of permits negatively affects business, employment, and tax base in coastal communities.
- Find that local residents disproportionately support local fish industry infrastructure (harbors, docks, and services supporting processors and fleet.)

Because of the limited entry constitutional amendment, a CPB program has a good chance of surviving **state** constitutional challenge.

Federal constitution

I think the federal constitution presents greater concerns. The primary hurdles are: (1) Privileges and Immunities Clause; and (2) Commerce Clause.

The basic concern is that a state statute cannot **discriminate** against **or** present impermissible **burdens** on residents of other states. Because federal Commerce Clause challenges have taken a "back seat" to Privileges and Immunities Clause challenges, I focus on the latter.

Toomer v. Witsell (1948) struck down a South Carolina statue requiring nonresidents to pay 100 times as much as residents for a shrimp fishing license. A state cannot discriminate where there is "no substantial reason" beyond "mere fact that they are citizens of another state."

Toomer: A "substantial reason" exists where a state shows that nonresidents constitute a "**peculiar source of the evil**" the statute seeks to address. But the state must **also** show a **reasonable relationship** between the discrimination and the "danger presented by nonresidents."

Hicklin v. Orbeck struck down an Alaska hiring preference statute because it was drawn too broadly to show a "reasonable relationship".

But the *City of Camden* (New Jersey) case (1984) held that a city ordinance favoring local hire might be allowed. **This is an important case.** Camden's ordinance required that 40% of employees and contractors on city-funded construction projects be local city residents.

The court said that where the city could show that nonresidents, **including other New Jersey residents who were not residents of Camden**, were a "source of the evil,"—i.e., spiraling local unemployment, eroding tax base—at which the ordinance was aimed, could survive challenge. *Camden* quoted *Toomey*: "States should have considerable leeway in analyzing local evil and prescribing appropriate cures".

Two other things in *Camden* are useful in defending a CPB program: (1) The court pointed out that the city's ordinance would **not completely foreclose nonresident** employment/contracting; 60% of the opportunities would remain unaffected. A CPB system would not remove nonresident's abilities to purchase limited entry permits. (2) The court revived "**state ownership**" as a factor. The Camden ordinance was limited to publicly funded city projects. Fish in Alaska's waters are a state resource recognized by courts.

The moral of the story is: don't tailor a statute to explicitly discriminate against non-Alaskans, *per se*; instead, affirmatively state the positive purposes, supported by findings, of granting local residents the exclusive ability to lease CPB-administered permits.

(Deborah Mercy photo.)

Federal Fisheries Management Policy and Legal Realities

Rachel Baker
NOAA National Marine Fisheries Service, Juneau, Alaska

I appreciate the opportunity to participate in the Fisheries Access for Alaska workshop. I agree with other speakers that maintaining access to fisheries is a critical issue for our state. My presentation focuses on federal fisheries management in Alaska from a legal and policy perspective.

Federally managed fisheries

To begin, federal fisheries are those from 3 to 200 miles offshore, in the Exclusive Economic Zone or EEZ. Fisheries in the EEZ are primarily managed under the authority of the Magnuson-Stevens Fishery Conservation and Management Act or the Magnuson-Stevens Act. This map shows the federal groundfish management areas that have been established in Alaska's EEZ, just to provide some context. The primary areas in the EEZ are the Bering Sea, Aleutian Islands, and Gulf of Alaska.

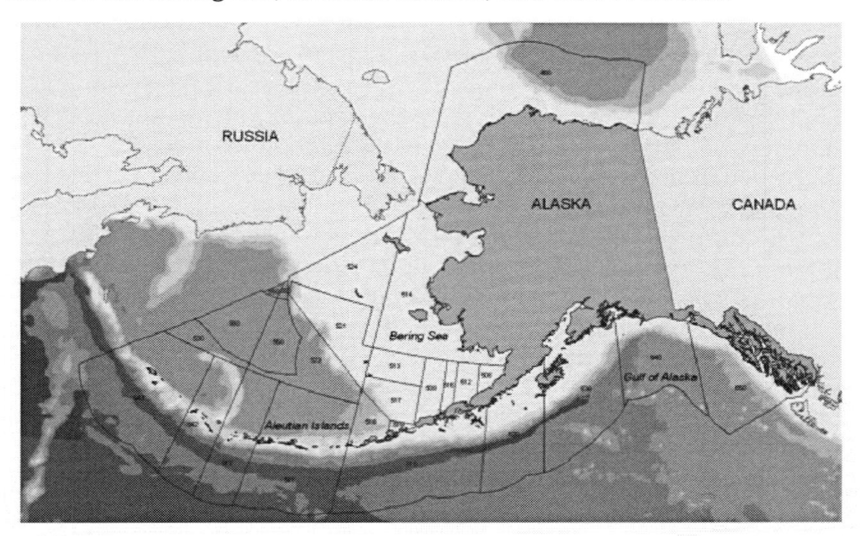

An earlier version of the Magnuson-Stevens Act was first authorized in 1976. The Act established the EEZ and the current management framework for federal fisheries in Alaska. Congress has made two significant revisions to the Magnuson-Stevens Act since then—first in 1996 with the passage of the Sustainable Fisheries Act, and then in 2007 with the Magnuson-Stevens Fishery Conservation and Management Reauthorization Act. These revisions contained a number of provisions that were drawn directly from fisheries management processes in Alaska, specifically for provisions related to limited access and catch share programs, which are discussed here.

The Magnuson-Stevens Act is unique among federal management statutes in that it designed a regionalized fishery management structure to allow regional, participatory governance by knowledgeable people with a stake in fishery management. The Act established the North Pacific Fishery Management Council for the Alaska EEZ as one of eight regional fishery management councils in the United States. The Council is the primary body responsible for developing management policy and programs for federal fisheries in Alaska.

North Pacific Fishery Management Council

The Magnuson-Stevens Act directs councils to have representation from coastal states, affected fishery stakeholders, and the relevant state and federal management agencies. The North Pacific Council has eleven voting members: seven fishing industry representatives and four management agency representatives. The seven fishing industry representatives are recommended by the governors of Alaska and Washington and appointed to the Council by the Secretary of Commerce.

Alaska has the majority of voting members on the Council: six out of eleven. The commissioner of the Alaska Department of Fish and Game has a permanent seat and the governor of Alaska recommends five Alaskans for rotating three-year terms to represent the fishing industry. The State of Washington has three voting members and the State of Oregon and the National Marine Fisheries Service (NMFS) each have one voting member.

The Council holds five public meetings each year, in addition to number of committee and workgroup meetings throughout the year on specific issues. The Council receives public comment for every item on its meeting agenda prior to taking action.

The Council's primary responsibility is development of fishery management plans (FMPs), or management policies for the fisheries. The Council is also responsible for the ongoing management of federal fisheries, which typically involves amendments to fishery FMPs or to fishery management regulations. In Alaska, the development and implementation of Council management policies is a coordinated effort among the Council, Alaska Department of Fish and Game, International Pacific Halibut Commission, and NMFS.

Fishery management plans

The Council has developed six FMPs for fisheries in Alaska's EEZ: Gulf of Alaska Groundfish, Bering Sea and Aleutian Islands Groundfish, Bering Sea and Aleutian Islands King and Tanner Crabs, Statewide Scallop Fishery, Salmon in the EEZ off Alaska, and fishery resources of the Arctic Management Area. Because fisheries and fishing fleets are dynamic, the Council is continuously amending these FMPs to ensure that management and regulations are consistent with its policy objectives. For example, each of the groundfish FMPs has been amended more than 100 times.

Major fisheries managed by the Council

This graphic shows the primary species that are managed in the federal fisheries. Several species of groundfish are managed under the groundfish FMPs in the Bering Sea and Aleutian Islands and the Gulf of Alaska, including pollock, Pacific cod, rockfish, sablefish, and flatfish.

The crab FMP defers most crab fishery management activities to the State of Alaska, including the opening and closing of fisheries and setting total allowable catches or guideline harvest levels for the fisheries.

Nine of the Bering Sea and Aleutian Islands crab fisheries are managed under the Crab Rationalization program developed by the Council, which allocates harvesting and processing privileges to specific fishery participants.

And finally, the Council is also involved in management of the directed halibut fisheries along with the International Pacific Halibut Commission. Management of Pacific halibut is governed by a treaty between the United States and Canada, and US fisheries are managed under the authority of the Northern Pacific Halibut Act of 1982. The Council established the halibut and sablefish individual fishing quota, or IFQ program under authorities provided by the Magnuson-Stevens Act and the Halibut Act.

Although the Council management structure is tailored for specific regional fisheries and intended to be responsive to changing conditions, all Council actions must be implemented by NMFS through a formal notice and comment rulemaking process in compliance with all Magnuson-Stevens Act requirements, the Administrative Procedure Act, and a number of other applicable laws. As part of this process, NMFS must review all FMPs, FMP amendments, and regulations developed by the Council to ensure that they are consistent with all requirements before approving them for implementation. The effect of this requirement is that all actions recommended by the Council undergo a thorough technical and legal review process within NMFS, NOAA, and the Department of Commerce before they can be implemented.

The requirement for NMFS to review and approve Council recommendations results in a longer process for development and implementation of federal fishery management actions compared to State of Alaska management actions. This is a source of frustration among many policy makers and stakeholders in our fisheries, and it directly impacts the Council's development of management actions and priorities. Our goal at NMFS is to participate in the Council process as early as possible to minimize the agency review time for actions recommended by the Council, but all Council actions must go through this process.

FMP amendment and regulatory process

The Council has an established process to analyze and get public input on management actions under consideration, as shown in the graphic below. Most actions are on the Council's agenda at least three times before the Council makes a recommendation for an FMP or regulatory amendment.

After the Council takes final action to recommend an FMP or regulatory amendment, it forwards the action to NMFS for review and approval. NMFS publishes a proposed rule for the action and accepts public comment on the proposed rule for a minimum of 30 days. After the public comment period closes, NMFS considers all comments

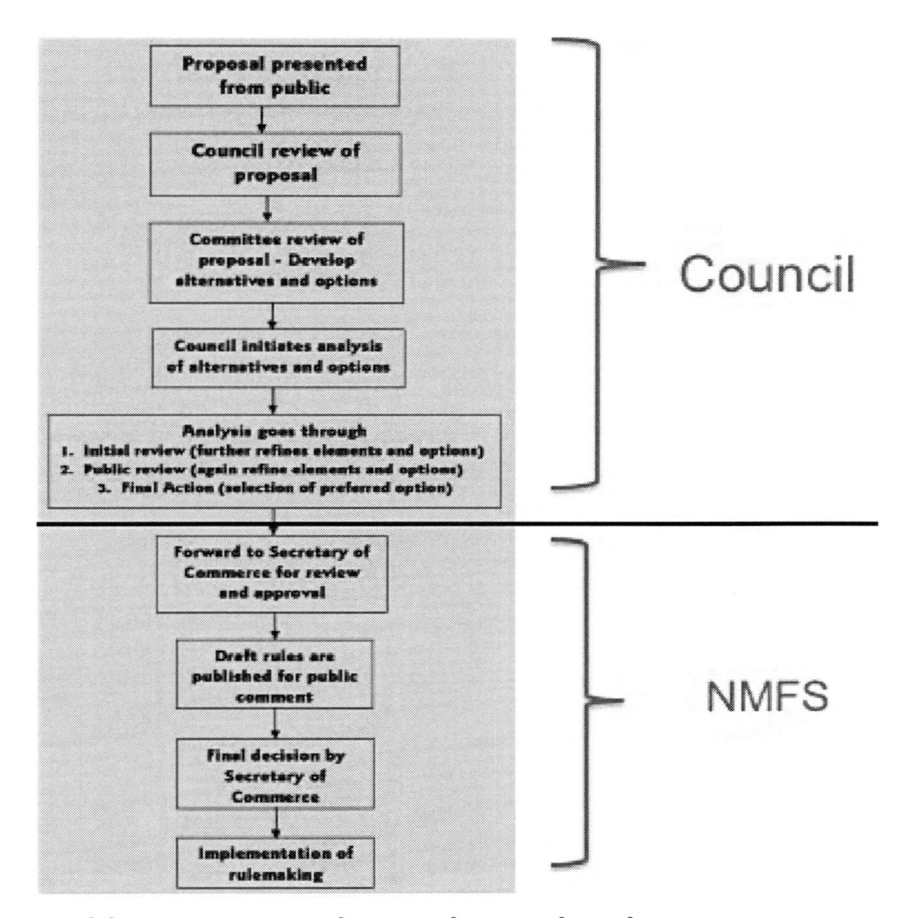

Fishery Management Plan amendment and regulatory process

received and prepares a final rule to implement the action. The final rule must respond to all of the comments received on the proposed rule. NMFS cannot implement the Council's action until the final rule is published and effective.

The Magnuson-Stevens Act also requires all FMP and regulatory amendments to comply with a number of conservation and management requirements, including the ten national standards codified in the Magnuson-Stevens Act. The standards are principles that promote sustainable fisheries management. The national standards require the Council to balance a number of considerations, including preventing overfishing, promoting achievement of optimum yield from the fisheries, economic efficiency, and fishery safety. Although the Council must consider all ten national standards in any action it recommends and

NMFS must determine that the action is consistent with all ten national standards, in practice, the Council typically focuses a management action to be particularly responsive to between two to four of the national standards. For example, past Council actions to reduce bycatch in specific fisheries were primarily focused on balancing national standard 9 requirements to minimize bycatch to the extent practicable with the requirements of national standard 1 to achieve, on a continuing basis, optimum yield from the target fisheries.

National standard 4

Among the ten national standards, two are particularly relevant to this discussion for maintaining local access to fisheries: national standard 4 and national standard 8. National standard 4 prevents management actions from discriminating between residents of different states. In practice, this generally means that the Council cannot establish fishery participation criteria or allocations based solely on participants' state of residence. National standard 4 also governs the allocation of fishing privileges, specifying that allocations must be fair and equitable and be established such that no person acquires an excessive shares of fishing privileges.

I think it is important to note that the Magnuson-Stevens Act does not define "fair and equitable" or "excessive shares." Each council has the responsibility and the flexibility to determine appropriate management measures for specific fisheries.

While national standard 4 is clear that management actions cannot discriminate between residents of different states, the Council must balance that directive with national standard 8, which highlights the importance of fishing communities. National standard 8 requires the Council to consider the importance of fishery resources to fishing communities for all of its management actions. National standard 8 also specifies that Council actions must provide for sustained participation of fishing communities and to the extent practicable, minimize adverse impacts on fishing communities.

Unlike some of the terms in national standard 4, the Magnuson-Stevens Act does define the term "fishing community" as "a community which is substantially dependent on or substantially engaged in the harvest or processing of fishery resources to meet social and economic needs, and includes fishing vessel owners, operators, and crew and United States fish processors that are based in such community." So that is a fairly comprehensive definition and generally includes the entities the Council has considered to be part of the fishing community in its management actions.

As you can see, the Magnuson-Stevens Act national standards explicitly address the allocation of fishing privileges and requires the

Council to consider the impacts of allocation on individual fishermen and on fishing communities. A number of Council management actions have focused primarily on national standard 8. For example, the Council established the Community Quota Entity Program, or CQE Program, to provide Gulf of Alaska coastal communities and the community of Adak with an opportunity to establish a nonprofit entity to purchase halibut and sablefish quota share on behalf of the community. The entity then leases the quota to community residents to be fished. I will provide a few details about the CQE Program from a policy and legal perspective.

In developing the CQE Program, the Council defined communities that were eligible based on specific criteria, including population, proximity to the fisheries, and historical participation in the fisheries. In approving the CQE Program and subsequent amendments to the program, the Council and NMFS have noted that the CQE Program is consistent with national standard 4 because while all of the CQE communities are in Alaska, both Alaska and non-Alaska communities were determined to be not eligible for the CQE Program. Therefore, the management action was not predicated upon any effort to discriminate between residents of different states. Furthermore, many of the benefits of the IFQ program were anticipated to continue to accrue to participants that live outside of CQE communities. Finally, the Council determined that the CQE Program was consistent with national standard 8 because it was intended to provide fishing opportunities for residents of fishery-dependent coastal communities, in addition to minimizing the adverse economic impacts of the costs of entering the current IFQ program for CQE Program fishery participants.

From a policy perspective, the important point is that the Council developed a community quota purchase program for specific Alaska communities that met all of the requirements of the Magnuson-Stevens Act. To do this, the Council identified very specific objectives, designed the program to meet those objectives, and received stakeholder support for the program.

I would note, however, that as the Council has continued to expand the CQE program to authorize community entities to hold additional privileges, such as groundfish License Limitation Program (LLP) licenses and charter halibut permits, some non-CQE participants in the IFQ fisheries have opposed these actions, asserting that they violate national standard 4 because the actions discriminate against non-Alaska residents and are not fair and equitable. I think it is an important policy reality that the Council is likely to receive negative feedback from some fishery participants when it considers actions that are intended to develop new access tools or expand existing tools for new entrants, residents of coastal communities, or community entities.

Limited access programs and limited access privilege programs in the Magnuson-Stevens Act

The Magnuson-Stevens Act gets even more specific in the provisions related to development of limited access programs, or limited entry programs, and especially for limited access privilege, or catch share programs, in which exclusive fishing privileges are allocated to individuals, cooperatives, or other types of entities. The 2007 reauthorization of the Magnuson-Stevens Act established section 303A, which specifically addresses catch share programs. Section 303A does not require the Council to consider implementation of a catch share program. The Council has the authority and the flexibility to determine if a catch share program is appropriate for any of the fisheries it manages.

If the Council does elect to establish a catch share program, section 303A specifies several requirements for the program. Primary among these requirements is that the program will help rebuild fish stocks, contribute to reducing capacity in the fishery, improve operational efficiency, promote fishing safety, and provide social and economic benefits to fishery participants. These objectives are consistent with those established for the catch share programs that have been implemented in Alaska.

Based in part on experience with catch share programs in Alaska, section 303A specifically addresses the allocation of fishing privileges. Section 303A requires the Council to make fair and equitable allocations of fishing privileges that consider the impacts of catch share allocations on the cultural and social framework of the fishery.

The Magnuson-Stevens Act also specifies that the Council should develop catch share program policies that (1) promote sustained participation of small owner-operated fishing vessels and fishing communities that depend on fisheries, and (2) establish measures to prevent excessive consolidation in the harvesting or processing sectors of the fishery.

Section 303A also specifies that, if appropriate, the Council should include measures that assist entry-level and small vessel owner-operators, captains, crew, and fishing communities through set-asides of harvesting allocations or economic assistance in the purchase of fishing privileges.

Finally, Section 303A established two specific types of entities that the Council may consider authorizing to use fishing privileges in a catch share program: fishing communities and regional fishery associations. I'll note that while the Magnuson-Stevens Act authorizes and provides specific requirements for the Council to establish fishing community and regional fishery association entities, the Council does not have to adopt these provisions as codified in the Magnuson-Stevens Act and may authorize other types of entities to use quota in a catch share program subject to requirements established by the Council.

But as I mentioned earlier in the discussion of the national standards, I think a policy reality is that the Council will have to consider the impacts on and perspectives of current fishery participants if it is considering establishing community-based entities to receive allocations of quota. Some fishery participants may perceive allocations to community entities as a "take away" from their fishing history and may not support including these types of provisions in management programs.

Alaska limited access and catch share programs

Many of the catch share program requirements and directions for policy considerations in the Magnuson-Stevens Act reauthorization of 2007 are based on Alaska's experience with limited access and catch share programs.

Several limited access and catch share programs implemented in Alaska are relevant to our discussion. These are the License Limitation Program for groundfish, implemented in 2000; the halibut and sablefish IFQ Program, implemented in 1995; the BSAI crab rationalization program, implemented in 2005; and the Central Gulf of Alaska Rockfish Program, implemented in 2012. Each program includes a number of tools that were intended to promote specific objectives within the programs. These tools are:

- Quota and vessel use caps to prevent excessive consolidation, maintain fleet diversity, and broadly distribute economic benefits.
- Provisions to promote or require active participation in the fishery.
- Exemptions from licensing requirements or separate allocations to small vessel/entry-level participants.
- Allocations of quota or licenses to community entities.
- Community quota purchase options.
- Delivery requirements to maintain historical distribution of economic benefits.
- Sideboards in other fisheries to prevent quota recipients from expanding into non–catch share fisheries.

Summary

From my perspective, the Magnuson-Stevens Act and the federal fisheries management system provide flexibility to tailor limited access or catch share programs to achieve specific objectives, such as new entry or allocations of quota to communities. A catch share is a tool to meet specified objectives. For example, if you identify new entry as an objective up front, the program can be designed specifically to promote that objective.

It is important to keep in mind that establishing and modifying federal management programs does take some time, since all actions must go through the Council process. As discussed, the Council has the authority to emphasize national standard 8 for management actions to benefit communities in close proximity to the fisheries, but during the development of such actions it must consider the national standard 4 requirement to prevent unlawful discrimination among residents of different states. This typically requires a balancing of interests and perspectives among the different stakeholders throughout the Council process.

Balancing

In general, the Magnuson-Stevens Act provides a very flexible framework under which a wide variety of management tools can be used to promote local access to fisheries, as long as the Council provides a rational basis for why it believes the tool will meet its program objectives. Often one the most challenging parts of developing programs or management tools is the prioritization of objectives. The Council must always consider a number of trade-offs when designing limited access or catch share programs.

In addition, the Council is continually challenged with balancing a number of competing stakeholder interests to develop a program that meets policy objectives and is supported by affected participants.

From my perspective, another challenge for the Council is to design management programs that use the power of markets to achieve specific economic and social objectives. Many of the access tools discussed here are actually introducing inefficiencies into a market-based system to meet a specified objective. Preventing excessive consolidation, for example, or limiting quota leasing to preserve existing business operations can reduce the economic benefits from a program and limit flexibility for fishery participants to react to changing conditions.

It is also important to remember that building these types of inefficiencies introduces management costs and creates additional burdens for fishermen. Specifically, tools implemented to promote or require active participation result in increased reporting burdens on fishery participants and are more costly for agencies to monitor. For example, the Council has implemented a number of active participation requirements to purchase halibut, sablefish, and crab quota by transfer. Fisherman wanting to purchase quota have to submit the required documentation to demonstrate active participation and NMFS must verify the documentation before a quota transfer can take place. While these active participation tools can be effective in meeting Council objectives for the program, it is important to analyze the impacts of increased reporting and monitoring costs and balance those costs with the anticipated benefits of including the tools in the catch share program.

(Deborah Mercy photo.)

Community Permit Banks

Representative Jonathan Kreiss-Tomkins
Alaska State Legislature, Juneau, Alaska

Alaska's commercial fisheries are a critical and sustainable source of employment, income, cultural identity, and state and local tax revenue for Alaska and its people. Over 30,000 commercial fishermen harvest over $6 billion worth of fish and shellfish in waters off Alaska each year. Many coastal communities have few alternative economic and job opportunities outside commercial fishing.

Alaskans want a vibrant and sustainable Alaska fishing industry supporting economically empowered and self-sufficient Alaska communities.

State-issued limited entry fishing permits, however, are leaving Alaska, particularly rural Alaska, and nonresident ownership is increasing. Between 1975 and 2014, Alaska's rural communities experienced a net loss of over 2,300 limited entry permits. And in the Bristol Bay salmon fishery, local permit ownership declined by 50% between 1975 and 2014.

Permits are getting more expensive—the value of Alaska's limited entry salmon permits has more than doubled in the last 15 years. The high cost of entry makes it difficult for new and young Alaska fishermen to enter Alaska fisheries. Young fishermen, especially fishermen from rural areas, lack the necessary capital and collateral to obtain loans. Even when funding is available, the risk new fishermen face is staggering. This "graying of the fleet" is a phenomenon that is well documented. In 1980, the average age of an Alaska fishery permit holder was just over 39 years; in 2014, the average age was nearly 50. Few alternative employment opportunities exist in these communities, so losing access means losing livelihood and ultimately losing community.

We want to allow communities to band together and create permit banks. New and young fishermen can access permits from permit banks as a stepping-stone before buying a permit outright and becoming independent fishermen. Just as one usually rents before buying a house, permit banks create the opportunity for new and young fishermen to lease a permit for a few years, make some mistakes, save some money, and get their legs solidly under them before making the major life commitment, and taking on tens or hundreds of thousands of dollars of debt, to buy a permit outright.

Just as the Alaska Division of Investments' Commercial Fishing Revolving Loan Fund has helped thousands of Alaskans enter Alaska fisheries, permit banks are a tool to help Alaskans access Alaska fisheries, and to empower coastal Alaska communities with greater economic self-sufficiency.

Permit banks aim to reverse the dual trends of the "graying of the fleet" and of permit out-migration from Alaska by creating a tool that improves Alaskans' access and opportunity in our fisheries.

Efforts to Regain Permits: Successes and Challenges in Bristol Bay

Alice Ruby and Robert Heyano
Bristol Bay Economic Development Corporation, Dillingham, Alaska

Thank you for the opportunity to present BBEDC's efforts. You can access the BBEDC Program Guide and Annual Report at www.bbedc.com. We provide these to all of our residents annually.

Introduction

BBEDC has invested millions of dollars directly or indirectly into the effort to reverse the flow of permits out of the region. Some of our programs are unique in the state. We spent a great deal of time developing the models because there weren't any that we could draw upon. We're very proud of our efforts.

A critical point that I would like to emphasize is that BBEDC's position has been that just putting a permit in a resident's hands is not enough. Our objective must be to assure successful resident fishing operations.

Part of the effort to create successful operations is to strike the balance of a commitment by the resident and an investment by BBEDC. We also design our programs to meet the requirements of our designation as an IRS 501(c) 4. Every BBEDC program has been designed to fill a need in the region and to demonstrate each participant's need for the benefits of that program.

Our programs include:

- **Permit Loan Program** (PLP) (described below)
- **Emergency Transfer Grant Program:** Provides eligible residents with financial assistance to obtain an Emergency Transfer permit. We discovered that this was critical in order for our younger/potential fishers to gain the experience, and sometimes the cash,

necessary to move into permit ownership. In 2014, BBEDC spent $226,000 and enabled 30 residents to hold and fish limited entry permits. Nine moved forward to file PLP applications at the end of the 2014 season.

- **Vessel Acquisition Program:** This works much like our Permit Loan Program to provide financial assistance as well as business counseling and training to help permit holders purchase a vessel.
- **Interest Rate Assistance:** Provides 4% interest assistance up to $4,000 for loans for permits, boats, and gear. In 2014 we spent $47,000 for about 35 participants.
- **Shore Fishery Lease Grant:** One-time grant of up to $800 toward obtaining a shore fishery lease. We have awarded three grants since this program was introduced in 2013.
- **Personal Finance Program:** This offers one-on-one financial counseling with the staff of our partner organization, Money Management International.
- **Technical Assistance Program:** One-on-one consultation with the staff of our partner organization, Alaska Business Development Center.
- **Vessel Upgrade Grant Program:** Grants of up to $17,500 for set net permit holders and up to $35,000 for drift permit holders for a variety of upgrades to boats and engines. Grants can also be combined with qualifying commercial fishing loans. In 2014 we provided 91 drifters/set netters with grants that exceeded a total of $1.2 million. An In-season Emergency Provision was added this year to provide financial assistance to fishers for equipment and repairs to an engine or drive train that is catastrophic in nature to the operation.
- **Quality Improvements Program:** Begun in 2003 and expanded in 2014 to all watershed communities. BBEDC spent just under $70,000 in 2014 to provide slush ice bags, insulated totes, padded deck mats, and flexible sheet foam hold insulation. At the end of 2014, the total historical participation totaled about 850 residents.
- **RSW Purchase Program:** Provides grants to assist with the purchase of refrigerated seawater system for existing vessel.
- **RSW Support Program:** Provides assistance to fishers with existing refrigerated seawater systems for consultation prior to and during installation of RSWs, spring tune-ups, and repairs of RSW systems. In 2014 BBEDC provided grants of up to $1,000 to 15 resident vessels.

- **Pre Season Advance Program:** Provides up to $5,000 to help resident fishermen pay pre-season expenses such as nets, insurance, license renewals, and other fishing necessities. The advance and a $25 application fee must be repaid by July 15 of the year in which it was advanced. In 2014, 39 participants received a total of $185,900.
- **Vocational Scholarship Program:** Provides scholarships for courses/training (where it results in a certificate) for training such as CPR, HACCP, RSW maintenance courses, vessel safety classes, etc.
- **Business of Fish:** Annual workshops in our larger communities provide an opportunity for fishers to meet with agencies and share commercial fishing information. Thanks to the participation and support of the Alaska Sea Grant Marine Advisory Program and the University of Alaska Fairbanks Bristol Bay Campus.

Permit Loan Program

I am most familiar with our Permit Loan Program and have been involved since we first introduced it in 2008. I will describe it in more detail to give you an idea of our development and delivery efforts.

With the help of Alaska Growth Capital and RedPoint Associates, BBEDC created the framework for our Permit Loan Program in about 2007 and rolled it out in 2008. We added revisions in 2012 to create the program as it is offered today. I should mention that work on a program to aid in reversing the flow of permits began over 10 years ago with a Bristol Bay Native Association effort to create a revolving loan fund. That did not prove to be feasible and BBEDC took over the research and development in about 2006.

Our goal for this program was to bring 7-15 participants into the program annually. Our research indicated that seven permits each year would contribute to slowing down the flow while 15 permits each year would begin to reverse the flow. Our initial efforts saw the following results. In 2008—2 permits, in 2009—3 permits, in 2010—3 permits and so on without accomplishing our target. It wasn't until 2013 that we began to accomplish our minimum requirements of seven permits in a year.

To date we have approved 33 residents for participation in the Permit Loan Program (there will be more joining in the early part of 2016). To date four participants have exited the program by moving out of the region, and as far as we know the four continue to commercial fish. All participants have loans with either the Alaska Commercial Fishing and Agriculture Bank (CFAB) or Alaska Division of Economic

Development (DED). Currently we have 27 applicants not yet entered into the program but in various stages of the application process.

The BBEDC Permit Loan Program provides assistance and benefits that enable eligible residents to obtain a permit (by way of a loan) and to successfully operate a commercial fishing enterprise. Applicants gain one or more of a menu of benefits provided in conjunction with a loan from either CFAB or DED. The program is available to all Bristol Bay watershed residents.

The menu of benefits is:

- Loan guarantee of 25% to 75%.
- Interest subsidy of up to 4.5% annually, up to $4,000 or actual interest due.
- Equity (sweat equity) assistance of up to 50% disbursed over an established period.
- Mandatory annual and regular financial counseling.
- Mandatory and regular business counseling.
- Down payment assistance of up to 95% of the down payment/ closing costs as determined by the lender.
- Repayment of all funds granted if the resident exits the program. Exit occurs if the resident does not meet the annual eligibility requirements for the duration of their loan term.

In order to stay in the program, the participants must demonstrate their eligibility on an annual basis:

- Maintain residency in the CDQ region.
- Annual participation as commercial fishers.
- Maintain the permit and loan in good standing.
- Participate in minimum annual training.
- Participate in minimum annual financial and business counseling. This is a very important part of our program and we depend heavily on assistance from our partners, the Alaska Business Development Center, and Money Management International.

The application process works as follows.

1. Resident submits an application. There are two stages to process the application

a. General eligibility: residency, fishing experience, no legal issues that put the permit at risk, has a market, can provide tax returns.

b. Financial analysis: We work with the individual to assess their financial situation to determine the menu of services that will be needed to assure the individual is lendable and they can financially operate their business. Based on our recent history, about half of the applicants will need to be diverted to our partners for help with financial planning or similar.

2. If approved by BBEDC, a loan application is prepared and submitted to the lender(s).

3. If approved, and once the permit transfer occurs, the resident permit holder signs a contract with BBEDC to commit to the terms of the program for the duration of their participation in the program (the terms of their loan).

To date, we have a cumulative commitment of about $1.7 million in loan guarantees. This is based on the principal balance of the loans at closing. The guarantees range from 25% to 75% of the principal balance. This commitment changes annually as principal balances decline and new participants enter the program.

To date we have paid:

- $494,000 in down payment grants.
- $16,000 in sweat equity, with a commitment to participants for additional payments.
- $19,000 in interest subsidy, with a commitment to participants for additional payments.

As we go forward and based on 7-15 permit holders added annually, we would expect to commit a cumulative annual amount of approximately $6 million in loan guarantees and we expect to pay out about $4 million in sweat equity and interest subsidy by 2023.

Challenges

There is never a lack of interested residents, but we are challenged to recruit/approve applicants for reasons that we have grouped as follows:

- **Risk aversion:** Even with our assistance, the financial commitment for an individual to buy into the fishery is substantial both for financing their entry and for the cost of operating a fishing business. Our programs are designed to try to help residents overcome some of that aversion.

- **Concern about the long term stability of the fishing industry.**
- **Social pressure:** We spent a generation telling our young people to "get a career" but forgot to let them know that commercial fishing can be a viable and respectable career.
- **Residents' ability to borrow money:** Many clients have no credit, poor credit, or legal issues such as child support or with the Internal Revenue Service. My staff has observed that much of this comes from lack of experience with credit, personal financial management, and/or the transition from subsistence economy to cash economy. We can help our clients but in many cases, this means the application process can be longer—as much as three years while they work with our partners.

The following paragraphs expand further on the challenges that both support the need for our Permit Loan Program and impact the effort to draw residents into the program.

We can put numbers behind the need for the Permit Loan Program by looking at the cost of living in rural Alaska. The numbers come from work done in 2009-2012 for BBEDC by Northern Economics. Food and gas costs in Dillingham are 150% of Anchorage costs. Electricity in Dillingham costs twice that in Anchorage. Food in Dillingham is higher than other surveyed communities in Alaska. Dillingham is a regional hub, so the prices there are cheaper than other communities in the region. Coastal communities outside of Dillingham have a 7% higher cost of living. Also important is that 72% of Bristol Bay watershed drift permit holders identified commercial fishing as the sole source of income. It is slightly different for setnet permit holders—70%. It is really interesting that other Alaskans who hold Bristol Bay drift permits say that 76% of their income is from other employment opportunities. And in the setnet fishery other Alaskans state that 86% of their income is from other employment opportunities. You can see some of the hurdles we are facing.

A significant challenge for drawing residents into the program is that the applicant, regardless of what BBEDC thinks about him or her, still has to be approved by CFAB or the State of Alaska in reference to a loan to purchase the permit. An applicant may have a poor credit record, lack of a credit rating, or insufficient financial resources to make the repayment schedules. Another very important challenge is this: not only do we want the permits to come back to the region, we want those permit holders to be successful in the region, which presents some problems in itself. Some of the younger residents can successfully walk through the hurdles and get the permit. We need to further educate them that fishing isn't a business you are going to go broke on. You can be successful. We just need to encourage them, and spend more time working the numbers with them to show where and how it is possible.

There can be a lack of salmon markets for new entries. We are blessed with large salmon runs in Bristol Bay, but that leads to processors not looking for new fishermen, which poses a problem for someone looking to start their own commercial fishing business. One of the causes was a consolidation of Bristol Bay processors, leading to a lack of markets.

Another challenge that impacts our ability to draw residents into the program is adverse Board of Fish regulatory changes, especially regulations that result in further consolidation of wealth. It requires considerable capital to be a beneficiary of such regulatory changes, which further deprives our resident fishermen. Essentially we have lost our shoulder seasons in Bristol Bay, basically a sockeye fishery. Years ago we were able to start fishing the first of June and target kings, then we fished sockeye, and in the fall we would fish pinks and silvers. Those fisheries are basically gone. It is not really a big money fishery and it's pretty slow. Watershed residents who live there probably participate at a higher rate than people who don't call Bristol Bay their home.

It must be emphasized that there are successes. This program is unique, and is not available in any other region. It was invented here and we have very strong support from the BBEDC board in continuing the program. We have strong support within the region. Even though we can argue that the numbers aren't sufficient, we have had a 100% success rate. We did not have anyone default, which is important. We've seen an increase in applications. BBEDC looks at this as a work in progress and they are willing to listen to suggestions for changes not only coming from board members but also from watershed residents, and to incorporate those changes. There have been many changes to this program since it first started, and they are all beneficial for us to reach our goal. This program would not be as successful as it is if we did not get support from CFAB and the State of Alaska, especially for financing permits. We've gotten an excellent response from those who lend.

We are always happy to answer questions and we encourage folks to contact us.

Breakout group discussion. (Deborah Mercy photo.)

Alaska Division of Economic Development Loan Programs

Jim Andersen and Britteny Cioni-Haywood
Alaska Division of Economic Development, Anchorage, Alaska

Britteny Cioni-Haywood is the Director of the Division of Economic Development (DED) and Jim Andersen is the Loan Manager. DED has offices located in Anchorage and Juneau. The Anchorage office processes loans from Cordova and west, while the Juneau office processes loans from Yakutat on down the panhandle. All loan servicing is done through the Juneau office.

The Division of Economic Development administers and services several distinct revolving loan funds for the Department of Commerce, Community, & Economic Development (DCCED). These loan funds were created to strengthen and support Alaskan ownership in certain industries, or to facilitate economic development in rural areas of the state.

These are the loan programs we currently have active:

- Rural Development Initiative Fund
- Small Business Economic Development Fund
- Alaska Microloan Revolving Loan Fund
- Alternative Energy Conservation Loan Fund
- Commercial Charter Fisheries Revolving Loan Fund
- Alaska Capstone Avionics Loan Program
- Commercial Fishing Revolving Loan Fund
- Community Quota Entity Revolving Loan Fund
- Mariculture Revolving Loan Fund

1. The goal of the **Rural Development Initiative Fund** is to provide private sector employment by financing the startup and expansion of businesses that will create significant long-term employment. The fund can be used for any small business purpose; for example, for starting other business opportunities during off fishing times or when there is a downturn in fishing to diversify income. This fund has been quite successful in providing meaningful long-term employment in a lot of communities. They tend to be smaller loans. The general requirements are:

- Loans may be made to a business in a community with a population of 5,000 or less that is not connected by road or rail to Anchorage or Fairbanks, or with a population of 2,000 or less that is connected by road or rail to Fairbanks or Anchorage.
- Loans may be made for working capital, equipment, construction, or other commercial purposes.
- Loans may not be made to pay costs that were incurred more than six months prior to receipt of loan application.
- Loans must result in the creation of new jobs or the retention of existing jobs in the eligible community.

The maximum term is 25 years, the loans allow $150,000 to one person and $300,000 for two or more people, interest rates are fixed at the time of loan approval, and the loans are adequately secured.

2. The goal of the **Small Business Economic Development Revolving Loan Fund** is to provide private sector employment by financing the startup and expansion of businesses that will create significant long-term employment. General requirements are:

- Eligibility includes all communities in Alaska.
- Loans in communities of 30,000 or more are available on a limited basis, depending on fund availability.
- Loans must result in the creation or retention of jobs in eligible areas.
- There is a non-public matching fund requirement of 1.5 to 1.

The maximum term is 20 years for fixed asset and five years for working capital, the maximum loan amount is $300,000, interest rates are fixed at the time of loan approval, loans are adequately secured, and purchased assets must be collateral.

3. The purpose of the **Alaska Microloan Revolving Loan Fund** is to promote economic development in Alaska by helping small businesses access needed capital. Microloan is a brand new program—only

a few years old. It has been fairly successful at providing financing for folks who can't get a bank loan. The general requirements are:

- Loans may be made for working capital, equipment, construction, or other commercial purposes for a business located in Alaska.
- Must be an Alaska resident for the 12 months preceding the date of application.
- Loans may not be made to pay costs that were incurred more than six months prior to receipt of loan application.
- Applicant may not have any child support arrearage.

Terms and conditions:

- Maximum loan amount is $35,000 to a person or up to $70,000 to two or more persons.
- Loan requests of $35,000 or more require a letter of denial from a financial institution, stating the reasons for denial, or confirmation that a loan from a financial institution is contingent on the applicant receiving a loan from the fund.
- Maximum loan term is 6 years.
- Interest rate fixed at the time of loan approval, contact us for current rates.
- All loans must be adequately secured.
- Applicant must commit a reasonable amount of non-state funds to the project.

4. The purpose of the **Alternative Energy Conservation Revolving Loan Fund** is to provide commercial business owners access to affordable loans for energy conservation, retrofitting projects, and installation of alternative energy systems

Alternative Energy Systems: A source of thermal, mechanical, or electrical energy that is not dependent on oil, gas, or nuclear fuel for the supply of energy for space heating and cooling, refrigeration and cold storage, electrical power, mechanical power, or heating of water.

Commercial Building: A building intended to be used for commercial purposes excluding residential structures, apartment complexes of less than five units, and single units within a condominium or cooperative complex.

Energy Conservation Improvement: Structural insulation, thermal windows and doors, weatherizing, heat exchangers, high efficiency furnace and boiler additions.

General Requirements:

- Loans may be for the purchase, construction, or installation of alternative energy systems or energy conservation improvements in commercial buildings.
- Must be an Alaska resident for the 12 months preceding the date of application.
- Loans may not be made to pay costs that were incurred more than four months prior to receipt of loan application.
- Loans must result in alternative energy production or energy conservation.
- Applicant may not have any child support arrearage.

Terms and conditions:

- Maximum loan amount is $50,000. Loan requests over $30,000 require a letter of denial from a financial institution, stating the reasons for denial, or confirmation that a loan from a financial institution is contingent on the applicant receiving a loan from the fund.
- Maximum loan term is 20 years.
- Interest rate fixed at the time of loan approval, contact us for current rates.
- All loans must be adequately secured, include a lien on real property, and the improvements financed

5. The **Alaska Capstone Avionics Loan Program** provides long-term, low interest loans for purchase and installation of Next Generation (NextGen) avionics equipment for aircraft that operate in Alaska. The program has been very successful, and has saved a lot of lives.

6. The purpose of the **Commercial Fishing Revolving Loan Fund** is to promote Alaskan ownership, the development of predominantly resident fisheries, and facilitate the continued maintenance of commercial fishing gear and vessels by providing long-term, low interest loans. This is the largest and oldest of our loan programs. In the last few years

with the upturn in fishing, the loan program is fully utilized—we are currently about 90% lent out. The general requirements of the program are:

Borrowers must:

- Be a resident for the last 2 years
- Not owe past due child support
- Be in compliance with IRS filing requirements.

Loan may be used for:

- Permit Purchase
- Vessel Purchase
- Product Quality Improvement
- Engine Fuel Efficiency
- Gear Purchase
- Vessel Upgrade
- Tax Obligation
- Purchase Quota Shares

Terms:

- Maximum 15 year term.

7. The **Community Quota Entity Revolving Loan Fund** provides long-term low interest loans to recognized Community Quota Entities for the purchase of halibut and sablefish fishing quota. The quota will be leased to resident fishermen to provide community access to the fisheries. The intent of the CQE program and the new fund are to reverse the out-migration of quota from rural areas.

8. The **Commercial Charter Fisheries Revolving Loan Fund** enables eligible Alaskans to finance charter halibut permits and helps retain and grow permit ownership by Alaskans under the system instituted by the National Marine Fisheries Service.

9. The **Mariculture Revolving Loan Fund** provides long-term loans for the development of Alaskan-owned mariculture operations and helps diversify economic opportunities in Alaska's coastal communities.

That is a rundown of the Alaska Division of Economic Development loan programs.

Peter Pan Cannery.

Financing Options for the Next Generation

Lea Klingert
Commercial Fishing and Agriculture Bank, Anchorage, Alaska

Alaska Commercial Fishing and Agriculture Bank

I would like to start with a brief description of the company I work for, the Alaska Commercial Fishing and Agriculture Bank or CFAB. CFAB is a private member-owned finance cooperative and with minor exceptions is limited to lending to Alaska residents. CFAB opened its doors in 1980. It is the product of legislation introduced in the late 1970s and is the only private entity in Alaska that can take a lien on an Alaska Limited Entry Permit. CFAB was created to fill what was at that time a lack of options for financing for the Alaska commercial fishing industry.

Prior to serving in my current position as CEO, I was a loan officer and also managed the Credit Department at CFAB. I have been with them going on 30 years and during this time I have had the opportunity to observe the evolution of the participants and the industry. As I have listened to previous presenters I have heard several references to "lack of capital" and "lack of financing" as a roadblock to obtaining or maintaining ownership of permits in rural Alaska. My experience would suggest otherwise.

The rationalization and consolidation of the Alaska fisheries over the years have resulted in their becoming more stable from a financing perspective and as a result there are many lenders ready and willing to lend to the industry. Alaska residents have far more options available to them than other states. My experience would suggest that we do not have a lack of financing options—if anything we may have a lack of qualified applicants.

This issue is not unique to the commercial fishing industry and is not meant to be derogatory. Obtaining credit for any reason should never be "easy." The consequences in this arena can be very severe.

No one benefits from obtaining credit they can neither afford nor are prepared to manage.

CFAB has been partnering as a lender with the Bristol Bay Economic Development Corporation and its watershed residents to help them prepare for and enter into their own fishing operations. One of the key components of the program is focusing on personal fiscal management. While this program may not have yet delivered the numbers that BBEDC would like see, it is my opinion that the program is successful and should serve as a model moving forward. This program not only assists in providing an avenue for entrance into the fisheries, it also provides education so the individuals obtain the business acumen to manage their business profitably and thereby remain in the industry long term.

There is a lot of focus on financing costs and the difficulty of obtaining a loan. While the tendency is to make this cheaper and easier, this approach generally has the unintended consequence of driving up the costs of permits. While interest rates rise and fall, the price one pays for the permit or boat is permanent. Interest in getting into the fisheries tends to peak when prices are climbing. This is not the best time to invest; the better option is to be prepared to enter when the industry is declining and prices are falling. Preparing future fishers to be ready to invest at this time is essential, and utilizing the program that BBEDC has in place, or one like it, will go a long way toward achieving this goal.

As Gunnar Knapp mentioned at this workshop, there is a distinct conflict between making the fisheries more profitable for the participants and serving the needs of the local communities. It is important we keep in mind that any reduction in the number of permits available for purchase generally increases the cost or value of the permit. This then puts the permits out of reach for a lot of the local residents.

Participation loan

A participation loan is outside the traditional options currently available. This type of loan allows the seller, or perhaps a family member not involved in the transaction, to participate in the loan. Basically the participant is a co-lender with CFAB. They own a portion of the loan and can thereby set the terms on their portion of the loan. This can be advantageous to both parties as it is assumed that the participant is more familiar with the borrower and will have more knowledge from which to determine the borrower's character and capacity for the loan. It also can provide the participant with a source of revenue.

The fisher and/or the seller bring assets, fishing experience and knowledge, and knowledge of the buyer and/or the community to the transaction. CFAB provides documentation expertise, loan servicing, lien perfection/permits, arms-length transaction, and collection of loan payments.

Transaction sequence

1. Fisher decides to sell.
2. Fisher sets sales price.
3. Fisher locates buyer.
4. Buyer prepares loan application and submits it to CFAB for review.
5. CFAB reviews application and determines their risk threshold.
6. Seller and CFAB work out the terms of the new loan/participation, taking into consideration the short/long term financial needs of seller/participant.
7. Buyer agrees to the terms offered by CFAB and seller/loan participant.
8. CFAB prepares loan documents, secures collateral, and disburses funds to seller.
9. Buyer makes payments to CFAB, and CFAB services loan and forwards payments to seller/participant.
10. Buyer makes payments until loan is paid in full at which time all collateral is released.

For example, if the sale price is $500,000, the buyer's down payment is $50,000, CFAB puts up 30% of the loan or $135,000, and the seller holds 70% of the loan or $315,000. The seller gets $185,000 cash at the time of sale.

Loan terms and payment

CFAB 30% loan = $135,000
CFAB loan terms: 20 years @ 7.50% (variable) = $14,242 per year.
Seller/loan participant 70% loan = $315,000
Seller loan terms: 20 years @ 6.00% fixed = $27,500 per year
(seller/participant sets interest rate and term)
Total annual payment = **$41,742**

Distribution of payments

CFAB payment:

Interest	$10,125.00
Principal	$ 3,117.00
Total	$14,242.00

Loan balance owing after payment = $131,883

Seller/loan participant payment:

Interest	$18,900.00
Principal	$ 8,600.00
Total	$27,500.00

Loan balance owing after payment = $306,400
Loan servicing fee = $393.75
Check to seller/loan participant = $27,106.25

Another financing option

The existing vessel owner can sell a percentage of the vessel/operation over time to the buyer/crewmember/family member. Those incremental purchases can be financed by a bank or paid by the buyer. If they are financed by a bank the seller must be agreeable since it is likely the item being purchased will be used as collateral to secure the loan. This can provide tax savings to the seller and provides a way for the seller to phase out of the business over time, as well as a way for the buyer/crewmember to phase in a little bit at a time. It has the distinct advantage of providing the buyer with time to run his own operation and develop a track record on his own, which will assist him in getting financing on his own down the road.

Beyond financing

I would like to emphasize that my experience suggests that credit is not the issue here, and that we should not only be concerned with Alaskans getting into the fisheries but with providing them the tools to operate a fiscally sound operation so they may remain. It is very important that we not only address the issue of getting and keeping the permits in the state. It is important that we prepare the new entrants to manage their business in a way that provides them with a reasonable chance at success. In my experience catching the fish is just the beginning and usually the easier part; managing your income to your lifestyle with a one-time payday in my experience has proven to be the more difficult part.

What can a fisherman do today to prepare?

- Start planning. It is never too early.
- Start saving.
- Take business classes.
- Take money management training, especially for getting through 12 months on a three-shot payday.

- Find a mentor.
- Research, analyze, ask questions.
- Attend the Alaska Young Fishermen's Summit.
- Emergency transfer a permit for experience.
- Remember, the fisherman you see fishing beside you today did not get there overnight.
- Talk to the fishers in your community and ask how they did.

Petersburg, Alaska.

Educational Entry Permits and Emergency Transfers

Bruce Twomley
Alaska Commercial Fisheries Entry Commission, Juneau, Alaska

I spent 10 years suing the state and federal government as a lawyer with Alaska Legal Services. I was fortunate to be one of the lawyers representing Alaska Native village plaintiffs in the *Molly Hootch* case (a class action to secure high schools in villages). Governor Jay Hammond appointed me to the Alaska Commercial Fisheries Entry Commission on Halloween of 1982.

Educational entry permits

CFEC has statutory authority to issue Educational Entry Permits to accredited educational institutions in Alaska, who are training young people who are at least of middle school age. CFEC has supported 10 such programs over the years.

A successful program that received favorable attention was run by the Cordova High School. The program had a classroom component and also placed the students on working vessels for the season. Students were able to get in two years of participation toward state loan eligibility. Some graduated to become real commercial fishers.

The McCann Treatment Center in Bethel is the only remaining program. Fourteen resident middle school students participate with two instructors and a skiff. McCann has a classroom component that teaches fish identification, fishing techniques, safety, the history of commercial fishing with emphasis on the Yukon-Kuskokwim area, and commercial fishing regulations. Jim Andersen told me that students in such a program can earn State loan eligibility.

Anyone interested in exploring Educational Entry Permits in detail should call the head of our Licensing Section, Yvonne Fink, at 907-790-6952. I would hope this opportunity might serve to offset the disengagement of youth from commercial fishing described by Courtney Carothers.

Temporary emergency transfers

CFEC is also authorized to issue temporary emergency transfers primarily to permit holders who are disabled from fishing. But there is an interesting way in which emergency transfers intersect with opportunities for getting and keeping limited entry permits in families.

Years ago I was seated next to Norm Van Vactor on a plane in Bristol Bay. He asked me if there was some way to put off transfers to young people in their early twenties. He observed that permit holders of that age did not always succeed and sometimes sold their permit to buyers willing to pay substantial sums. He thought if such transfers could be put off until individuals were at least 25 years old, chances for success would improve. (This was an amazing insight, because the conversation took place years before anyone discovered that young people must be at least 25 before the frontal lobes of their brains mature and their judgment improves.)

When I got back to the office, we faced a request for an emergency transfer by a father who held a Bristol Bay drift permit and was permanently disabled from fishing. He had already been issued his allotted emergency transfers. The father wanted to permanently transfer the permit to his son, but a transfer at the time was impracticable, because the son was in treatment. We found authority in the *Ostrosky* case (which upheld the constitutionality of free transferability) to grant the emergency transfer, which we continued to do until the son achieved the capacity to function as a permit holder in his father's judgment.

This experience also prompted CFEC to modify our regulation governing emergency transfers. Under 20 AAC 05.1740(i)(2)(C), even after an individual has used up their allotted emergency transfers, an Extraordinary Circumstances Provision includes the opportunity to get additional emergency transfers if the permit holder has a plan to permanently transfer the permit, but cannot fulfill that plan at the time because of an obstacle (for example, the transferee lacks capacity at the time). So CFEC has some ways to buy time for a permit holder to make a thoughtful decision and not feel unnecessary pressure to do a permanent transfer.

60-day cooling off period

Another device that can help ensure that a permit holder makes a sound decision to transfer the permit is our statutory 60-day cooling off period before a transfer can take place. I once traveled to a jail in Anchorage at the request of my friend Jerry Liboff to meet the son of a Bristol Bay family. The son had entered an agreement to transfer a Bristol Bay drift permit to someone from California. We were within the 60-day cooling off period, and permanent transfer papers had not been filed. The son cooperated and signed a form withdrawing his 60-day notice of intent to Transfer, which I took back and filed at the commission. By timely withdrawing his notice of intent, the son (without any penalty under our statute) nullified his contract with the individual from California and later transferred the permit to a family member.

Thus CFEC has some tools to avoid intemperate transfers and to support family transfers. Additionally, we have employed emergency transfers, whenever doing so would help avoid the loss of an entry permit in the face of claims by the Internal Revenue Service, the Child Support Division, and foreclosures by one of the two authorized loan programs.

Countering federal demands

Some of you remember CFEC's more than a decade-long struggle with the IRS in their attempts to cease and force the sale of limited entry permits. IRS often targeted rural Alaska permit holders. That struggle culminated in our decision refusing an IRS demand to transfer the Southeast seine permit held by Francis Carle, a 61-year-old Alaska Native fisherman from Hydaburg. The IRS did not sue, and we achieved a stalemate.

Something similar is playing out now. A Togiak woman who held a set net permit became disabled. She transferred her permit by gift to her sister, who continued to use the permit for the benefit of their family. When the woman applied for a Social Security Income disability, she was disqualified because she had failed to get the dollar value of the permit during her transfer. The public assistance worker treated her permit as a resource and counted its value against her eligibility for SSI disability. We worked with a great lawyer, Tina Reigh, when she was at Alaska Legal Services in Dillingham. (She is now a magistrate judge.) I testified at the hearing before the Chief Administrative Law Judge in Anchorage, who ruled that the permit should not have been counted as a resource, and the woman was entitled to her SSI disability payments from the time they should have been awarded.

These decisions are not published, and therefore are not widely known. If you know someone whose permit is interfering with their eligibility for public assistance, you are welcome to a copy of the opinion, and you should refer them to Alaska Legal Services.

Q&A: Seller maintaining security

At the end of another presentation, Adelheid Herrmann asked about a seller of an entry permit maintaining a security interest in the permit to secure payments. The Limited Entry Act generally prohibits that. However, years ago, we helped CFAB change their statute so that an individual transferring a permit could participate in a loan by CFAB which would be secured by the permit. This accomplishes what Adelheid asked about, under current law. Lea Klingert described this opportunity in detail during her presentation.

Q&A: Senior permits

We have another question from Adelheid: Why did CFEC oppose a change of law to authorize senior citizens to lease their entry permits to family members? My recollection was that the Department of Law discouraged the idea, because it would grant a privilege to a limited group of people (seniors and family members) and create a risk that people outside of the select group could sue to either strike down the practice or expand the practice to other less desirable beneficiaries. The policy underlying the law that prohibits leasing is to maintain the independence of fishermen and avoid putting fishermen in a position where they can be exploited.

If anyone wishes to talk, Bruce Twomley's direct line is 907-790-6944.

Board Regulations That Encourage Local, Small Boat Participation

Kurt Iverson
Alaska Department of Fish and Game, Commercial Fisheries Division, Juneau, Alaska

Introduction

I am a program coordinator and analyst with the Commercial Fisheries Division of ADFG. My purpose here is to provide a brief overview of Board of Fisheries regulations that might encourage access to fisheries, or otherwise allocate fisheries toward small boat, less-capitalized fishing operations.

About the regulations

The Board of Fisheries has implemented a host of regulations across the spectrum of fisheries that essentially provide what economists often call "input controls" on fishing effort and on fishing capacity. For example, how many fisheries exist today that don't have gear limits?

Commonly, the input control regulations serve multiple functions:

1. They are designed to level the playing field in a competitive fishery, and therefore by their nature the regulations often skew harvesting allocations toward less capitalized fishing operations. The regulations also modestly hold down overcapitalization of fishing operations. The best of the regulations, however, also serve other legitimate functions, such as:

2. They serve conservation by slowing the pace of the fishery, which can help managers more effectively achieve their harvest goals.

3. They may also benefit enforcement (e.g., fishing in daylight hours) and encourage more orderly fisheries.

The Board of Fisheries has to work within established statutory and constitutional constraints. For example:

1. A Board regulation cannot run afoul of the Alaska constitutional guarantees of equal access and common use of resources. For example, a "locals-only" fishery cannot be implemented.

2. Generally, the Board cannot allocate fisheries resources within a gear type in a management area. This came out of the Alaska Supreme Court's *Grunert* decision.

3. *Hebert* is another example of an Alaska Supreme Court decision that helped define the limits of the Board of Fisheries' authority. In that decision, the Court upheld the Board's regulations for superexclusive registration in the Westward herring fisheries. Although superexclusive registration was acknowledged by the Court to primarily benefit local, small-boat fishermen, the regulations applied equally to all, so no unjust discrimination was found.

In addition to passing tests of legality, regulations should also be:

- Clearly enforceable.
- Straightforward and plainly understandable.
- Implemented by ADFG in an effective manner.
- Not conflict with other management or conservation goals.

The regulations are often controversial. Most allocation regulations are controversial because people are addressing their own best interests. Typically, persons who fish more aggressively and/or who have greater access to capital oppose regulations that constrain fishing capacity and skew harvesting opportunities away from their capacity and possibly toward persons who reside in economically disadvantaged locations or who are participating in less capitalized fishing operations.

Regulations that encourage access to fisheries (by constraining fishing power and restricting the consolidation of fishing capacity) are not always economically efficient. However, that is a balance we seek to achieve as a society. For example, the most efficient harvesting of salmon would be to take the fish in fish traps. But instead, the Board of Fisheries often seeks to support and enhance small-boat, owner-operator fishing businesses, and disallow excessive consolidation of fishing effort.

A brief summary of Board of Fishery Regulations that serve to enhance local, small-boat operations

Exclusive district registration

Salmon net registration areas

An individual or boat can participate in only one salmon net fishery statewide. Forms of this BOF regulation existed before statehood. CFEC has complementary regulations that help administer the BOF regulations.

Exclusive/superexclusive registration

Exclusive registration means that once a fisherman or vessel has registered for a particular management area, he or she can't register for any other area designated as exclusive or superexclusive. Note that in some Pacific cod fisheries, ADFG as been granted the authority to suspend this regulation in order to harvest remaining quota toward the end of the season. **Superexclusive** registration means that once a fisherman or vessel registers in an area with superexclusive designation, he or she can't fish anywhere else.

District registration

In the Bristol Bay salmon drift gillnet fishery, there is a 48-hour stand-down period for vessels or permit holders who seek to move from one fishing district to another. Moreover, the Bristol Bay Togiak District is effectively a superexclusive district within the Bristol Bay fishery. Another example occurs is in the Cook Inlet salmon set gillnet fishery, where permit holders can only fish one district for the entire year.

Vessel length limits

Vessel length limits are relatively common in state regulations. Following are some examples: For salmon seine fisheries, the vessel length limit is 58 feet. For the salmon fisheries in Bristol Bay, the vessel length limit is 32 feet. For the Pacific cod and Tanner crab fisheries in Chignik and the South Alaska Peninsula, the vessel length limit is 58 feet. And for the Adak (Area O) state-waters crab fishery, the vessel length limit is less than or equal to 60 feet.

Gear limits

Net gear and pot gear limits are more common than not. An exception to pot gear limits is the parallel fisheries for Pacific cod. Other examples of gear limits include prohibitions on power shakers in the Westward herring fishery. Another variation is that the number of divers operating a sea cucumber dive boat is limited to a maximum of two, with a maximum of three onboard. On scallop boats, no mechanical shuckers are allowed, and no more than 12 crewmen can work on the boats.

Fishing time

Fishing for sea cucumbers in Southeast Alaska is restricted to weekly, daylight-only dive openings, and for sea urchins the dive fishery openings are spread out to take place over a minimum of 16 weeks of openings within the season. For Pacific cod and Tanner crab, fishing is restricted to daylight hours in Westward areas.

Dedicated quotas

The Gulf of Alaska Pacific cod jig fisheries, which are often thought of as entry-level fisheries, are allocated a share of Pacific cod quotas in various management areas. Furthermore, if the jig quotas have not been taken at a specific point in time late in the season, they are normally rolled into the quotas for the pot cod fisheries. In some areas of the Gulf of Alaska, Pacific cod quotas are allocated by vessel size; for example, in Kodiak vessels over 58 feet are allocated a maximum 25% of the quota.

Trip limits—maximum harvest on board a vessel at any one time

The state-waters scallop fishery management plan includes the option for trip limits that can be imposed on vessels if they are needed to effectively manage for harvest limits. The Prince William Sound pollock trawl fishery has a 300,000 lb per trip limit, and the Aleutians Pacific cod trip poundage limit is 150,000 lb per day.

Dual permit operations

Dual permit regulations allow two permit holders to fish in tandem on one boat, and they are thereby allowed to fish a greater amount of gear than what is allowed for boats with only one permit holder. Regulations allow dual permit operations in the Bristol Bay and Cook Inlet salmon drift gillnet fisheries, and in the Southeast Alaska herring gillnet fishery.

Dual permit regulations are controversial. There are two principal arguments, pro and con:

- Pro: An individual can enter a fishery without the necessary investment of a boat and/or gear, which are substantial barriers. An individual can also be provided with an "apprenticeship" or "break-in period" while owning a permit.

- Con: Dual operations might tend to be formed by more highly capitalized operations that can more effectively fish the extra gear and/or purchase a second permit to put in a family member's name.

Understanding the Western Alaska Community Development Quota Program

Larry Cotter
Aleutian Pribilof Island Community Development Association, Juneau, Alaska

The Community Development Quota (CDQ) Program provides economically disadvantaged communities in western Alaska with the opportunity to generate capital to develop stable local economies based on the fishing industry. CDQ allocates to eligible communities a percentage of all Bering Sea and Aleutian Islands quotas for groundfish, halibut, and crab. The Aleutian Pribilof Island Community Development Association (APICDA) is one of six CDQ groups.

The CDQ program was created by the North Pacific Fishery Management Council (NPFMC) in 1992 as a result of the inshore-offshore battles, where offshore factory trawlers and the shoreside processors were discussing who was going to get what. It was initiated by factory trawlers going into the Gulf of Alaska and taking entire pollock quota in about six weeks. This left the Kodiak community on the beach for the remainder of the year. So off we went into allocations.

A total of 65 villages are authorized under section 305(i)(1)(D) of the Magnuson-Stevens Fishery Conservation and Management Act (Magnuson-Stevens Act) to participate in the program. Each CDQ entity is federally recognized as an independent nonprofit organization with a separate board of directors and internal governance process.

There are six CDQ groups (see map). APICDA is a 501c3 group, and the other five are 501c4s. The CDQ program was supposed to sunset in three years. But it never did sunset—it continued to be expanded. Initially pollock was the quota allocated, at 7.5%. When the American Fisheries Act went into effect in 1998 the quotas were expanded from 7.5% to 10%. The same thing occurred when crab rationalization went into effect—the quotas were similarly expanded to 10%.

Some people wonder why the CDQ program is only in the Bering Sea and Aleutian Islands. In the beginning the first motion I made to establish the CDQ program was to make it statewide in all Alaska rural communities. The Makah tribe from Washington was present and they protested that they should be part of the program as well because they had a relationship with Alaska. The general counsel stepped forward and said if we made it exclusive to Alaska that would violate the national standards because we would be discriminating against residents of other states. The obvious next question was: if we made it for only *some* Alaskans would that be OK? They said that would be OK. So it became a Bering Sea Aleutian Islands program.

CDQ groups are blessed to have access to a lot of capital which makes things possible that are not possible for other groups in other regions. We use the money in a lot of ways. All the groups are focused to the extent that they can on helping, assisting, and encouraging their residents to participate in fisheries. That means in our regions providing loans to buy IFQs, boats, gear, etc.

In our region every community that has a year-round processing plant is at least stable. Every community that does not is in some level of decline, some more rapidly that others. People want to live in their home communities. If you are going to buy access to the seafood industry, you really need to have a seafood processing operation. That is the only way to stabilize and provide opportunities for the future. It is not easy. We have spent $22 million in False Pass in the last three years and we are still short $6 million. Next we will need to ask for another $20 million to expand our plant in Atka. Then we will need to get into St George and it just keeps going. That is an important component.

Can the CDQ program be replicated in other parts of Alaska? We are blessed in the Bering Sea given the amount of quota that is available, 2 million metric tons. You have something to work with. You don't have that size of a resource in the Gulf of Alaska. Pulling it into a CDQ program would be very difficult if not impossible. The size of the CDQ that results is not going to be enough to be worth the pain of getting there. I think that is the reality.

Fifteen years ago Southeast Alaska did not have a sea urchin fishery. I was a consultant then, and I was hired to help bring the fishery about. We worked with the Alaska Department of Fish and Game, and had surveys done. We could see that a fishery was soon going to develop. We knew exactly what was going to happen—a gold rush from folks down south. We would lose control, limited entry would kick in, and the familiar pattern would follow. So we developed legislation that would take sea urchin quota and put it into a quasi CDQ program, where it would be divided by different areas in Southeast and communities would be the ones invited to bid. They would be encouraged to work with local fishermen and local processors to put together a bid, and part of the bid was they would have to pay for the management of the fishery as well. It was a great program, with quota awarded to the common property owners which in this case was Alaskans. But it got messed up politically and when the bill came out it awarded based on the highest bidder, unfortunately. Hence it was never implemented.

But it was feasible then and will be feasible again in the future. One of the things to contemplate is: what is coming in the future? We used to have a huge shrimp fishery in the Gulf of Alaska. Now shrimp are gone but they could come back. Before they come back is the opportune time to grab the resource that does not yet exist but will. Fashion the program around that. Same with crab in the Gulf of Alaska. We need to be creative, think forward, grab opportunities when they exist, see in advance that there will be opportunities in fisheries tomorrow.

(Deborah Mercy photo.)

The Success Story of the Norton Sound Red King Crab Fishery

Tyler Rhodes
Norton Sound Economic Development Corporation, Nome, Alaska

Through the efforts of regional fishermen, both subsistence and commercial, the Norton Sound red king crab fishery evolved from one that was dominated by outside interests to one that supports a resident fleet and economy. While this story is presented by the Norton Sound Economic Development Corporation (NSEDC), the northernmost Community Development Quota (CDQ) group, it is a tale about the fishery as a whole.

Today the fishery is 100% harvested by residents of the Norton Sound region. The fishery provides opportunities for residents and fishermen, including seasonal employment in a Nome-based plant, tender boats, and support industries. The fishery is also biologically successful. The crab stock is healthy, growing, and thriving. A small-boat fleet and sound management have allowed stocks to rebuild and grow. All these elements were not the case during the first few years of the fishery. At its inception, indeed, it seemed the fishery was on another track.

Most of the fishermen participating in today's Norton Sound red king crab fishery are from Nome, Golovin, Elim, Shaktoolik, and Unalakleet. NSEDC's processing plant is located in Nome. This red king crab fishery is unique as Alaska's only commercial red king crab fishery to take place in the summer. It is primarily conducted derby-style from early July to mid-August. While the majority of the harvest takes place in the summer, the fishery also has a winter component, which occurs through the shore-fast sea ice.

1976-1981—Early commercial fishery

The commercial fishery was initially pushed for by Nome-area salmon fishermen who were also subsistence crab fishermen. Their desire to diversify helped lead to the creation of the commercial fishery. A survey was completed in 1976, and in 1977 an exploratory fishery harvested

500,000 pounds. In the next few years, annual harvests ratcheted up to 2 million to 3 million pounds.

In the early years (1977-1993) the fishery was dominated by outside fishermen who effectively blocked out locals. The level of capitalization needed to compete with the vessels coming into Norton Sound to fish was beyond what locals could do. Additionally, the fishery was prosecuted with catcher-processors and mothership fleets that would not buy from locals. The high harvest rates under the management regime at the time also threatened the stability of the stock.

All this precipitated a loud outcry from subsistence users in the region, who traditionally fished in the winter through the ice. After the first three years of the summer commercial fishery, the subsistence catch rate became severely low. Subsistence users blamed the exploding commercial fishery.

1981-1992—Decreased harvest level

Plummeting subsistence catches led to a survey to quantify what the harvest had been prior to the commercial fishery. The study was released in 1981, accompanied by a call for a moratorium on the summer commercial fishery in Norton Sound. Locals wanted regulations to provide for sufficient catch rates for winter subsistence and commercial fisheries. The moratorium did not occur, but in 1982 a provision essentially cut the commercial harvest level in half. The level had been 40% of legal males, which was then divided to 20% in summer and 20% in winter. The winter commercial harvest at the time never approached its 20% limit, effectively keeping the overall commercial harvest to the 20% cap.

This regulation change may be the move that saved Norton Sound's red king crab fishery from the fate that befell many of the fisheries to the south, that is, collapse or depressed to the point of slow recovery. The regulation cut harvest levels so the stock damage would not continue, allowing crabs to survive and rebuild in the following years. Subsistence users and local commercial operators both benefited from this in the long term. But in 1982 locals viewed it as too little too late, because they still could not get into the fishery.

During the 10-year period after the regulation change, from 1982 to 1992, the fishery was less attractive to larger vessels but they were not cut out altogether. The Norton Sound crab season, nestled between the Bristol Bay salmon season and the Bering Sea winter crab fisheries, made for a good opportunity to test crews and vessels.

The number of participating large boats varied from 3 to 27, depending on the year. This wide variation made it difficult to manage the fishery. In some years the same emergency order would announce both the opening and closure. There was little precision in management due to the ability of the fleet to overharvest very quickly. The fishery was unpredictable, unstable, and not economically productive for participants. Management varied from fairly good on low participation years to poor on high participation years.

The crab stocks also saw ups and downs. While there was a large recruitment event in 1985 and 1987, from pre-fishery parents, this was the last strong recruitment for the next decade—no males survived to terminal molt until the mid-1990s.

1993—Superexclusive fishery

In 1993 Norton Sound residents and NSEDC lobbied for a superexclusive fishery. This would mean that if a fisherman fished king crab in Norton Sound he or she could not fish it anywhere else during that year. The Board of Fish passed it that year, and in 1994 the North Pacific Fishery Management Council gave designation for the superexclusive fishery. That is what turned the tide to bring the fishery back to the locals. These regulation changes allowed the small-boat fishery and local processing to be established.

Before the superexclusive designation, there was an average of 125 pots per boat with the highest number at 400. Harvests in 1978-1981 were all over a million pounds. Almost 3 million pounds were taken in 1979 by 34 boats. Floating processors (catcher-processors or processors with motherships) were the norm. Management was very challenging—three boats were sometimes out in one season, and 30 boats the next.

After the superexclusive designation, pot limits were put in place. The season stretched from one month to two months. The largest boats were in the 35-foot range and many were smaller. This led to more precise management with a longer season and control of the harvest amount.

NSEDC

In mid-1990s the fishery was weak and still suffering from the aftereffects of large-boat overfishing. The buyers of the time were from outside the region and brought outside fishers to whom they guaranteed markets. Their limited processing capability put limits on local fishermen. To assist locals, NSEDC established a crab-buying operation to stabilize and introduce competitive pricing to the buying operations in the region. NSEDC's regional seafood operations, Norton Sound Seafood Products (NSSP), served Norton Sound and Yukon fishers first, and eventually limited buying to only them. However, a few fishers chose to sell to the other markets separately from NSSP.

NSEDC has played a large role in supporting both the fishery and fishermen. NSSP has run crab processing operations since the early

1990s, providing a market for resident crab fishermen. In 1994 NSSP processed 36% of the harvest; in 1995 they processed 98% and this has continued ever since. NSEDC has also invested in infrastructure to support the resident fishing fleet, from a seafood and freezer plant and harbor improvements to a fleet of tender vessels.

The Norton Sound Seafood Center opened with a modern processing plant in Nome in 2002. NSSP is a locally run and staffed company that employs the residents of Norton Sound communities, which strengthens the local economy. NSSP also has offered unique fishing opportunities, such as the winter, through-the-ice crab fishery, and has allowed the local fleet to diversify to halibut and cod fishing. This has provided economic stability through fishery diversification and season extension.

NSEDC also supports fishermen through a no/low-interest loan program for gear and vessels. In 2000 only seven of the seventeen vessels that fished the Norton Sound red king crab fishery were owned by Norton Sound residents. In 2001 a specific program was created for vessels intended for the crab fishery, aimed at turning that around. Over 11 years that program provided $3 million in loans to 51 fishermen.

On the regulatory side, a sustained harvest management strategy was put into place in 2012, with the harvest rate set at a sliding rate of 0, 7, 13, and 15%, depending on abundance. Since 2000, the participation rate in the fishery has remained fairly constant at 20 to 25 boats. The harvest has also remained fairly constant in the 300,000 to 400,000 pound range (summer only).

Since 2000 the crab population has gradually been increasing to support a stable and healthy crab fishery. This is due to a number of factors: responsible management, local cooperation and participation in management, and support of NSEDC through programs to support fishermen and seafood plant operations. Because resident fishermen make up the user group for Norton Sound crab, the region participates in management to the benefit of the fishery.

Acknowledgment

I would like to give credit to Charlie Lean, who provided a great deal of information for this presentation. He was an Alaska Department of Fish and Game fishery management biologist in Norton Sound for over 20 years before he joined NSEDC in the Fisheries Research and Development division. He continues to work for NSEDC, advising on fisheries issues.

CQE Program Description

Duncan Fields
Shoreside Consulting, Anchorage, Alaska

CQE Program

In 2002, the North Pacific Fishery Management Council recommended revisions to Individual Fishing Quota (IFQ) Program regulations and policy, to allow a nonprofit entity to hold quota share (QS) on behalf of residents of specific rural communities located adjacent to the Gulf of Alaska coast. In 2004, the National Marine Fisheries Service (NMFS) implemented the Council's recommendations as Amendment 66 to the Fishery Management Plan for the groundfish of the Gulf of Alaska. Amendment 66 implemented the Community Quota Entity Program (CQE Program) to allow these communities to form nonprofit corporations called CQEs to transfer and hold catcher vessel QS under the IFQ Program. Gulf of Alaska CQEs that transfer and hold QS on behalf of an eligible community may lease the resulting annual IFQ to fishermen who are residents of the community.

The Gulf of Alaska CQE Program was developed to allow a distinct set of small, remote coastal communities in Southeast and Southcentral Alaska to transfer and hold halibut and sablefish QS for use by community residents in order to help minimize adverse economic impacts of the IFQ Program on such communities and provide for the sustained participation of the communities in the IFQ fisheries. This program structure creates a permanent asset for the community to use. The structure promotes community access to QS to generate participation in, and fishery revenues from, the commercial halibut and sablefish fisheries.

Since the inception of the IFQ Program, many residents of Alaska's smaller remote coastal communities who held QS have transferred their QS to non-community residents or moved out of the smaller coastal communities. As a result, the number of resident QS holders has declined substantially in most of the Gulf of Alaska communities with IFQ Program participants. This transfer of halibut and sablefish QS and the associated fishing effort from the smaller remote coastal communities has limited the ability of residents to locally purchase or lease QS and has reduced the diversity of fisheries to which fishermen in remote coastal communities have access. The ability of fishermen in a remote coastal community to purchase QS or maintain existing QS may be limited by a variety of factors both shared among and unique to each community. Although the specific causes for decreasing QS holdings in a specific community may vary, the net effect is overall lower participation by residents of these communities in the halibut and sablefish IFQ fisheries. The substantial decline in the number of resident QS holders and the total amount of QS held by residents of remote coastal communities may have aggravated unemployment and related social and economic conditions in those communities.

The CQE Program includes a number of management provisions that originated from the IFQ Program structure and affect the use of CQE-held QS and the annual IFQ derived from the QS. The provisions include management area and vessel size category designations for QS, QS use caps, and QS blocks. Under some of these provisions, a CQE has the same privileges and is held to the same limitations as individual QS holders in the IFQ fishery. For example, CQE-held QS is subject to the same IFQ regulatory area use cap that applies to non–CQE held QS. In other instances, the CQE is subject to less restrictive provisions than individual, non–CQE QS holders. For example, a community resident leasing IFQ from a CQE may fish the IFQ derived from QS assigned to a larger vessel size category on a smaller size category of catcher vessel. In other instances, the CQE must operate under more restrictive provisions than individual, non–CQE QS holders, in part to protect existing QS holders and preserve "entry-level" opportunities for new entrants.

When the CQE Program was developed, the Council and NMFS were concerned that CQEs would try to acquire as much of the most affordable QS as they were allowed to hold and that gains in CQE holdings could reflect losses of QS holdings among residents of the same CQE communities. The Council and NMFS were also concerned that CQEs might have greater access to capital than would individuals, so they could buy up blocks of QS that are most in demand by non-CQE fishermen with small operations. The Council and NMFS determined it was appropriate to restrict CQEs from transferring or holding small blocks of QS to preserve fishing opportunities for new entrants in certain IFQ regulatory areas.

CQEs participating in the CQE Program have made little progress toward reaching the regulatory limits on the maximum amount of QS that may be transferred or IFQ that may be harvested. Since implementation of the CQE program in 2004, only two of the 45 communities eligible for the CQE program have formed CQEs, transferred QS, and harvested the resulting IFQ. These two CQEs do not hold sablefish QS. Based on a review of the CQE Program in 2010, the Council determined that lack of participation in the CQE Program can be attributed to limited availability of QS for transfer, increased market prices for halibut and sablefish QS, and limited viable options for financing QS transfer.

Since the CQE Program began, NMFS has implemented regulations that authorize the allocation of limited access fishing privileges for the guided sport halibut fishery and the Gulf of Alaska groundfish fishery for Pacific cod, to be allotted to select communities that are eligible to form a CQE. For the guided sport halibut fishery, the Council recommended and NMFS authorized certain communities in Southeast Alaska and Southcentral Alaska, Areas 2C and 3A, to request and receive a limited number of charter halibut permits, and designate a charter operator to use a community charter halibut permit to participate in the charter halibut fisheries.

The Council and NMFS wanted to enhance access to the groundfish and halibut fisheries and generate revenues for communities. Furthermore, the Council and NMFS wanted to provide for direct participation by individuals residing in, or operating out of, CQE communities.

Actual experience with CQEs

I have worked closely with two CQEs that have purchased halibut quota—Cape Barnabas, Inc. (Old Harbor) and Ouzinkie Community Holding, Inc. Now there are 33 registered CQEs out of 46 eligible communities. Twenty-one CQE communities have acquired halibut charter permits. It is not possible to determine how many of these are used. Also, in Areas 3A and 3B four CQEs have acquired 30 groundfish License Limitation Programs, but it appears that only five of these have been fished.

There are a number of draft bylaws and organizational documents. My experience is to balance simplicity with equity and/or objectivity. For example, regulations require that for the CQE the entity submit "detailed statement describing the procedures that will be used to determine the distribution of IFQ to residents of each community represented by that CQE." I worked with others and the Gulf of Alaska Coastal Communities Coalition to create rather elaborate distribution formulas. We had multiple criteria for vessel and gear ownership, experience, fishing plan, and discounts for quota already owned. In addition we set up opportunities for both large boats and small boats to share "pools" of quota. Each category was awarded points on an objective basis.

My experience is that the community CQE board moves toward equity and uses the allocation structure as a general guideline. I've seen the CQE boards often give more quota to a community member with more need or decide on an equal distribution of available quota regardless of the points earned.

Recommendation

My advice is to develop the objective scoring criteria but provide flexibility for the board to also add "equity" points, especially with the lower amounts of quota. **Experience has also taught that a successful CQE will need to try to secure a "backup" fishing vessel or individual, especially late in the season, in case something goes wrong with one of the transferees!**

Financing CQE quota

I've worked with three entities to finance quota. (1) At a **traditional bank** the requirement is 20% down, with a five year note, a higher interest rate, and a required loan guarantee. (2) With a **social objectives lender** there is a seven year term and lower yearly payments with a balloon, the interest rate is below market, and a guarantor is still required. One fisherman was able to refinance but with a higher interest rate. The social objectives lender required "best fishing practices." This is easy for the CQEs—usually small boat fishermen employ best practices anyway. (3) With the State of Alaska loan program, the term is longer, even up to 25 years. The down payment can be as low as 5%, there is some flexibility in a poor year, and the program has a pre-qualification option. I've generally had a good experience with the state loan program but the CQE still needs to find the down payment.

The Kodiak Archipelago Rural Regional Leadership Forum—"No Community Left Behind": A Project of the Kodiak Island Housing Authority

Roberta J. Townsend Vennel
Kodiak Archipelago Rural Regional Leadership Forum, Kodiak, Alaska

The Kodiak archipelago is a group of islands located in the Gulf of Alaska approximately 252 air miles south of Anchorage, Alaska, and is home to the regional hub of the city of Kodiak and six small, primarily Alaska Native communities. Kodiak Island is the largest island in the archipelago and the second largest island in the United States. The Kodiak Archipelago Rural Regional Leadership Forum (the Forum) is a consortium of five of these small coastal maritime communities, specifically Akhiok, Larsen Bay, Old Harbor, Ouzinkie, and Port Lions.

Community planning efforts supported by the Kodiak Island Housing Authority (KIHA) from 2004 to 2007 indicated that our region's small, coastal communities were struggling to have an impact on issues that were adversely affecting their quality of life. Issues such as very limited employment opportunities and limited to poor public education were responses received in community surveys as to why families were making the choice to leave their home community and move primarily to Anchorage. KIHA developed the Forum in direct response to the need for small coastal communities to work together on identified issues of common concern and to develop a unified voice. The Forum held its first session in fall 2005 and has continued to meet in late September, January, and April of each year. Each Forum session lasts for at least two-and-a-half days. Hallmarks of its successful approach include:

- Rural community leaders are invited to participate and the cost of travel, housing, and meals is supported through KIHA.

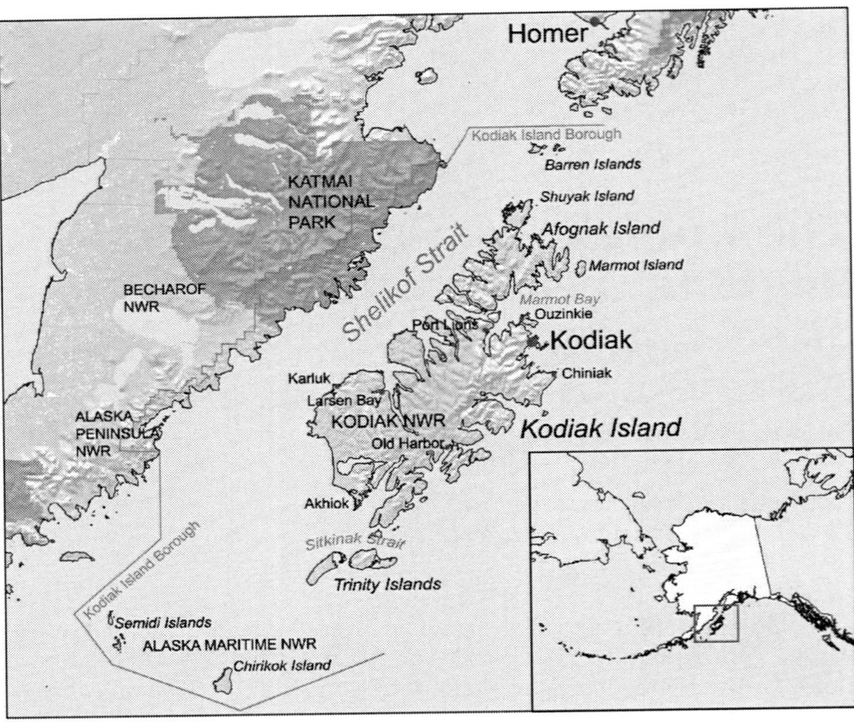

Map source: https://commons.wikimedia.org/wiki/File:Kodiakislandmap.png

- The Forum has no elected officials such as a chair or vice-chair. A facilitator is utilized to develop the Forum agenda, invite government and agency partners and presenters, and keep the sessions on track.
- The Forum has adopted a set of practices that respects the equality of each voice, practices respectful listening, and works to develop a consensus of how to move forward on addressing issues of concern.
- The Forum has no full-time staff. KIHA provides financial support for the facilitator and rural leadership travel as well as administrative support. Kodiak College, University of Alaska Anchorage, provides the Forum's venue and some administrative support while the Forum is in session. Forum government and agency partners contribute time and pay their own travel expenses to come and work with the Forum. And the region's Alaska Native Claims Settlement Act (ANCSA) corporations support board and staff members' travel and time to participate in the Forum.

The Forum focuses its efforts in three major ways. It:

- **Provides critical information** that speaks to issues identified as part of its agenda, and then develops follow-up work sessions to determine next steps.

- **Leverages existing resources** through the identification and development of Forum partners (governmental, agency, nonprofits, and others) and we work to bring partner resources to bear on identified priorities.

- **Supports strategic planning processes** by providing the planning venue and implementation support when required and when budget allows.

Over the past 11 years Forum participation has widened greatly. We now have an average of 36 community leaders participating at each Forum. Forum-identified priorities are moved forward by bringing together rural leadership with elected officials at the state and borough level, agency partners such as the Alaska Energy Authority, the Southwest Alaska Municipal Conference, regional and local law enforcement, Kodiak Island Borough School District Board of Education and administrators, leadership in the arena of fisheries policy, and others. As participation has widened, the ability for our rural communities to collaborate in a way that effects positive change has increased dra-

Former State of Alaska Commissioner of Education Dr. Larry LeDoux, sharing his rural education experience at the January 2016 Kodiak Archipelago Rural Regional Leadership Forum. Photo courtesy of Koniag Inc.

matically. This has resulted in widespread recognition of the Forum's effectiveness and stabilization of its financial support base.

The Forum's current agenda

As Forum participation has grown so has the breadth of its agenda. As demonstrated by the table below, the Forum is currently supporting a number of major projects in areas such as energy and food security. The Forum provided the venue and assisted with logistics and planning efforts in support of Phase I and II of the Kodiak Regional Energy Plan and is working with the Southwest Alaska Municipal Conference and the Kodiak Area Native Association to transition planning efforts into actual projects. In spring 2015 the Forum and over 15 partners embarked on a highly ambitious planning project to address often stated concerns and issues expressed at the Forum regarding rural and Alaska Native education in the region. The Kodiak Rural Regional and Alaska Native Student Education Strategic Plan (KRANS) was rolled out in September 2015 in a work session that included school board members, administrators, Alaska Native leadership, advisory school board members, and rural community leaders. Since September the working group has been active in implementing many of the plan's objectives. The Forum was also the incubator for the project titled The Small Tribes of the Kodiak Archipelago—Economic Stability through Food Security, a $1.2 million Administration for Native American Grant that will establish pilot farms in four of our communities.

Regional Government Collaborative Worksession	•Legislative Priorites and Capital Projects •Metal Debris and Household Hazardous Waste Cleanup and Removal
Transportation	•Kodiak Rural Regional Transportation Plan •Marine Transportation System Planning
Energy	•Kodiak Regional Energy Plan—planning venue for Phases I and II
Education	•Kodiak Rural Regional and Alaska Native Education Strategic Plan and Implementation
Food Security	•ANA $1.2 million grant to establish pilot farms in Larsen Bay, Old Harbor, Ouzinkie, and Port Lions
Community Safety	•Work sessions on critical topics
Fisheries	•Regional identification of issues, strategies, and goals •Intensive work sessions to move agenda forward

Forum engagement with fisheries policies

We are northern coastal maritime peoples whose identity is deeply connected to the sea. The Kodiak archipelago is renowned for the quality of its marine resources and is the center of fishing activities for the eastern Gulf of Alaska. With large populations of all five Pacific salmon species returning to the archipelago each year, the City of Kodiak's port is consistently in the top ten of US ports for value of fish delivered by the archipelago's fishing fleet. Kodiak's commercial fishing industry is also one of Alaska's oldest, dating back to the early 1800s when the Russians built the first salmon cannery in the village of Karluk, on the south end of Kodiak Island. Our communities are part of the last remaining active maritime subsistence continuum in North America. This continuum stretches from Alaska's Aleutian Islands to the Bering Sea and northern Canada and includes the Sugpiaq, Unangax^ (Aleut), Yu'pik, Chu'pik, Inupiaq, and Inuit peoples. For over 7,000 years the Sugpiaq/Alutiiq of Kodiak Island have depended on Kodiak's marine resources, first as a traditional hunting and gathering peoples, and in more recent years practicing a combination of commercial fishing and subsistence (personal use) harvest.

The archipelago's small coastal communities have demonstrated amazing resiliency over their 7,000 year history. They have survived wholesale destruction as a people through invasion and disease and natural disasters such as the Katmai eruption of 1912 and the 1964 Great Alaskan Earthquake that destroyed working waterfronts and fish processing facilities around the archipelago. Up until the early 1970s, Kodiak's waters and its marine resources supported thriving local Native village economies. Families enjoyed relative prosperity in their remote homeland as they practiced a lifestyle based on a combination of commercial and subsistence fishing.

But, beginning in the 1970s there were growing indications that something was very wrong. And that our once resilient communities were quickly losing ground. Indicators identified through community planning processes that our communities are in deep trouble include:

- Small boat harbors, the economic hub of a coastal community, have contracted from a viable small working fleet to just 1-2 working boats over the past 40 years.

- The community leadership pool and school enrollments have declined as families are forced to leave to seek economic opportunities elsewhere.

- An observed increase in behavioral health and community safety issues including suicide, drug and alcohol use, and public safety issues.

- State of Alaska salmon limited entry, a type of fisheries privatization, was implemented in the 1970s followed by federal privatization programs such as individual fishing quotas (IFQs) for halibut and sablefish.

There was a sense that the privatization of these key marine resources was at the root of the decline of our communities. But in the early years of the Forum there was no research to connect privatization with the crisis our communities are facing. There was only a very deeply seated sense of pain and loss.

However, the social science research of Dr. Courtney Carothers opened a door for the Forum to begin to engage with fisheries policy in a meaningful way. Dr. Carothers did her dissertation research in three of our Kodiak communities, Larsen Bay, Old Harbor, and Ouzinkie, and through her presentations at the Forum beginning in 2007, we began to get our arms around the story of "what happened?". Dr. Carothers' research showed clear linkages between privatization of both state and federal fisheries and declining access in small, coastal communities that led to a loss of economic opportunity for young men wanting to enter the fishery. Highly emotional Forum presentations and discussions, where older fishermen and community leaders came to the realization that fisheries' policies were contributing to the economic crises experienced in their communities, led to the Forum being able to craft effective work sessions. Through working with Forum partners such as then Kodiak Island Borough (KIB) Mayor Jerome Selby and KIB fisheries consultant Denby Lloyd, the Forum over several work sessions crafted a one-page working draft (see below) of fisheries issues, goals, and strategies to guide its work.

<table>
<tr><td>Kodiak Archipelago Rural Regional Leadership Forum
Working Draft Fisheries Issues, Goals and Strategies- 2013</td></tr>
<tr><td>ISSUES</td></tr>
<tr><td>Loss of Human Right to fish leads to a significant decline in access to fisheries by local communities and an increase in controlling interests by outside third parties. This loss of access impacts our communities in four major areas. Including:
• Degradation of Sugpiaq and community traditions that have been based on marine resources for at least 7,000 years. As the number of fishermen decline, valuable knowledge regarding the local environment and resources is lost.
• Loss of community food security. The ability to obtain and share subsistence foods is directly related to the number of commercial fishermen in each community as our commercial fishermen also fish for subsistence resources for large, extended families and community elders.
• Decline in available capital. As access to fisheries declines for our coastal communities, so does access to capital to purchase resources such as permits, IFQ's, boats and gear.
• Limiting opportunities. All of this contributes to limiting opportunity for our community members, particularly our young people. Without the ability to work on boats owned within the extended family, our youth have no ability to learn the skills developed over generations. This has led to a loss of work ethic. Their attitude is changing from "work for it" to "give it to me".</td></tr>
<tr><td>Fisheries management is currently not supporting the long-term sustainability of our marine resources for our small coastal communities. For example, regulations currently allow for fishing practices that can cause habitat destruction and large by-catch of non-targeted species such as halibut and king salmon that are important community subsistence resources.</td></tr>
<tr><td>GOALS AND STRATEGIES</td></tr>
<tr><td>Establish Fisheries Policies that support local communities:
➢ All State and Federal policies must be reviewed and modified to recognize a human, cultural and community's right to fish.
➢ All future policy development must be supportive of viable coastal communities.
➢ Fisheries policy must be viewed as a holistic systems, and policies must include consideration of the impacts on local economies, culture and heritage, teaching and learning between the generations, and community health and well-being.
➢ Food and livelihood sovereignty must be respected.</td></tr>
<tr><td>Making what we currently have work. Don't lose any more ground.
1. Work to reduce the cost of entry for our local fishermen.
2. Maximize effectiveness of Community Quota Entities.
3. Encourage our young people to participate in the cod jigging fishery before that closes.</td></tr>
<tr><td>Improve fisheries management, particularly in regard to critical subsistence species.
• Establish community exclusion zones (LAMP's) similar to what was done in Sitka.</td></tr>
</table>

Forum deliverables—fisheries policy

Now our Forum work sessions are focused in regard to fisheries policy so that we can tell our story and add our voices to the policy process. Fisheries is now a major part of each Forum's agenda and in any given Forum you'll find us addressing one or more of four major areas as identified through our planning process. These include:

Telling the story. The Forum works to share information on the impacts of current fisheries policies, including venues such as this conference. The Forum also invites researchers to share research design to provide input on current efforts that gather data to continue to articulate what is happening and why.

Information exchange. We invite advocacy groups, policy makers, fishermen, and others to participate in Forum work sessions so that information can be shared and utilized to assist community leadership in understanding the issues and policy making process.

Establish goals. Through the Forum we work toward consensus of our rural, regional community goals for fisheries policy.

Supporting goals. The Forum works with its partners to draft resolutions and work with cities and tribes to provide testimony to policy-making bodies such as the North Pacific Fishery Management Council. We also work to communicate our goals to other leaders such as our Alaska State Legislators and KIB assembly members.

Forum partners—fisheries policy

Over the past seven years, the Forum has worked with a group of key partners in regard to fisheries policy and social research. They contribute their time and energy on behalf of small coastal communities to assist us in dealing with the complexities of fisheries management and its policy makers. Our Forum partners include:

Dr. Alisha Drabek and Amy Steffien. While Alisha was the director of the Alutiiq Museum she and Amy developed a 7,000-year outline of the history of fishing in the archipelago based on the archaeology record that has proved helpful in demonstrating both the breadth of species utilized pre-contact as well as the resiliency of our people.

Dr. Courtney Carothers. Courtney is currently an associate professor with the School of Fisheries and Ocean Sciences at the University of Alaska Fairbanks. She has researched and published extensively on the impacts of limited entry and other privatization programs. Courtney has been part of the Forum since its inception and is regarded as part of our Forum family.

Duncan Fields. Duncan comes from a long-standing Kodiak fishing family and through his company, Shoreside Consulting, has assisted several of our coastal communities in accessing policy-making bodies. Duncan served as a member of the North Pacific Fishery Management Council for nine years. His breadth of experience both as a policy maker and as a fisherman make him a highly valued member of the Forum.

Representative Jonathan Kreiss-Tomkins. Jonathan is working on a critical access bill with the Alaska State Legislature that would create regional permit banks for salmon limited entry permits. This could create a pathway for young fishermen to enter the state's salmon fishery.

Darren Muller. Darren is on the board of the Gulf of Alaska Coastal Community Coalition and is also a board member of the Ouzinkie Native Corporation.

Theresa Peterson. Theresa assists the Forum in her role as Kodiak outreach coordinator for the Alaska Marine Conservation Council. She works with our rural leadership in developing a deeper understanding of process and actions to take to influence policy makers.

Takeaways from the Forum experience

There are four major takeaways that could translate the Forum's regional approach to a statewide organization of small coastal communities and other stakeholders who have experienced negative impacts from fisheries privatization programs. These include:

Articulate the story. Let's work together to identify critical information gaps and work to get relevant research funded. One major question in light of the State of Alaska's recent study of the state of food security in Alaska is—what are the long-term impacts on our state's food security under current state and federal fisheries policies?

Propose strategic solutions. It seems so often that smaller stakeholders and coastal communities are just "picking the crumbs off the table" left behind by larger, well-financed and politically sophisticated players. How do we change the current paradigm?

Build the constituency as a statewide issue and priority. Let's refocus the dialog so that others who may not live in coastal communities or are active fishermen see themselves as stakeholders. Issues such as the state's food security, survival of indigenous knowledge systems, and impacts on rural hubs such as Kodiak could be further defined and communicated in such a way that this becomes an Alaska priority.

Organize a statewide approach. Conferences such as this one could be the beginning.

In closing, Kodiak's marine resources have culturally defined our coastal communities for over 7,000 years but today we find that our very livelihood has been taken away. We are working together at the Forum to help take it back. We have a very long way to go to recover what has been lost over the past 40 years. Our hope is by working together and supporting communities to bring back fisheries access fisherman by fisherman, boat by boat, we can begin to rebuild and once again look forward to viable, sustainable coastal communities.

For further information please contact:

Kodiak Archipelago Rural Regional Leadership Forum
Robbie Townsend Vennel
907-299-6185
kodiakruralleadership@gmail.com

Rachel Donkersloot coordinates a breakout discussion. (Deborah Mercy photo.)

Creative Financing Mechanisms at Haa Aaní

Ed Davis
Haa Aaní Community Development Fund, Inc., Juneau, Alaska

The Haa Aaní Community Development Fund is a Native community financial institution based in Southeast Alaska. The impetus behind the creation of the Haa Aaní Community Development Fund was that we saw an outmigration of young people from rural communities. Our board at Sealaska Corporation felt compelled to create this economic development arm focused on rural community development.

The Haa Aaní Community Development Fund was established in 2012 by Haa Aaní, LLC, as a 501c3 with mission-driven initiatives to promote economic resiliency in Southeast Alaska. It was seeded with $500,000. Current assets are worth $3.3 million, about 80% of our loans have been distributed to rural communities, and the loans deployed have created and supported 42 jobs in five different communities. We are proud of that, but we are not done—we will continue our focus. Our funding partners include the Ramuson Foundation, The Nature Conservancy, US Department of Agriculture, US Treasury CDFI (Community Development Financial Institutions) Fund, and the Moore Foundation.

Our objectives are to develop and support sustainable business models, to build collaborative networks to leverage economic development resources in the region, and to provide financing/technical services to small businesses throughout the region.

A CDFI is a specialized mission-driven financial institution that serves underserved or economically distressed communities. CDFIs provide affordable access to financial products and services to market niches that are underserved by traditional financial institutions. Since 1996, the US Treasury CDFI Fund has awarded $1.5 billion, helping build a network of nearly 1,000 CDFIs across the United States. Alaska has six CDFIs. Since 1996, Alaska CDFIs have received over $17 million from the CDFI Fund.

The Haa Aaní Community Development Fund wants to improve the economic state of our rural communities. Economic data show 19 of 32 Southeast rural communities have experienced negative job growth since 2003. The hardest hit communities experienced a combined reduction of over 750 jobs (24% or 1 in 4 jobs). All rural jobs are down 8%. When I presented data to my board, one of the comments was that the communities with negative growth are largely traditional fishing communities. The average unemployment rate in rural Southeast Alaska is 11%, more than twice the average of urban areas of Juneau, Sitka, and Ketchikan.

Permits have significantly declined in Sealaska's traditional fishing communities. The average age of permit holders is 49.7 years, up 10 years since 1980. Between 1980 and 2013, Alaska residents under the age of 40 holding permits fell from 38.5% to 17.3% of the total permits. The smallest communities have been hit the hardest; for example, Angoon is down 91% from 20 years ago.

	1993 CFEC permits	**2003 CFEC permits**	**2013 CFEC permits**	**20-year % change**
Angoon	160	79	15	–91%
Craig	480	344	273	–43%
Haines	305	185	172	–44%
Hoonah	266	176	127	–52%
Hydaburg	105	56	39	–37%
Kake	180	93	59	–7%
Klawock	115	62	66	–43%
Pelican	188	81	47	–75%
Wrangell	665	484	384	–42%
Yakutat	282	230	252	–11%

CFEC = Alaska Commercial Fishing Entry Commission

The Haa Aaní Community Development Fund has several programs.

- The Loan Program offers affordable and flexible loan packages.
- The Path to Prosperity (P2P) regional business plan development competition has attracted 105 participants representing 18 communities and 12 industries, over the past three years.
- The Sustainable Southeast Partnerships (SSP) is a regional network of partners promoting community sustainability in four focus areas (natural resources, energy, food security, and economic development) with community and regional catalysts in over six communities.
- Workforce Development is building capacity with public and private partners such as the Hoonah Native Forest Partnership.

The Haa Aaní Community Development Fund offers two products—financial capital and technical services. **Financial capital** focuses on entrepreneurs interested in growing their business in their communities. We work with existing community partners and businesses that offer job sustainability. In addition we help entrepreneurs focus on triple bottom-line metrics to sustain their business model, and help establish and build credit to be bankable. **Technical services** focuses on community support events, business plan support and competition, and workforce development. This arm builds capacity at the community level. It is a source of capital and technical services for entrepreneurs, provides a lead role for community economic strategic plan development, and facilitates community partnerships to build capacity. The business plan competition attracted 26 applicants, with 11 finalists, demonstrating a strong entrepreneurial spirit in the Southeast Alaska region.

In summary, the Haa Aaní Community Development Fund provides many fisheries opportunities. As a CDFI, we have a team and partnership base to service the industry. We have established partnerships with the State of Alaska revolving loan fund, Alaska Sustainable Fisheries Trust, The Nature Conservancy, and others. We have flexible loan packages. Our goal is to help build capacity and promote resiliency.

Quinhagak, Alaska.

Alaska Sustainable Fisheries Trust Local Fish Fund—Investing in Alaska's Fishing Future

Linda Behnken
Alaska Longline Fishermen's Association, Sitka, Alaska

With few alternative income sources, Alaska's coastal residents depend on access to healthy fisheries. Limited access programs have dramatically increased the cost of entry to Alaska fisheries, creating economic barriers for rural Alaskans and reducing opportunities in rural communities. Fishing is integral to the identity of coastal people; losing access means losing a way of life and, ultimately, losing communities.

Driving the rising cost of entry to Alaska's fisheries is the escalating value of limited entry permits and individual fishing quotas. For example, the cost of entry to the halibut/sablefish quota share programs has increased 600% since the program was implemented in 1995. Rural residents have limited access to capital and often few employment alternatives to fishing, rendering these entry costs and risks all but insurmountable. Retiring fishermen care about the next generation of fishermen, but have significant investment to recoup and are looking for flexible exit strategies that also meet their retirement needs.

The Alaska Sustainable Fisheries Trust (ASFT) seeks to restore and retain fishery access for coastal Alaskans. The mission of the ASFT is to strengthen Alaska's fishing communities and marine resources through research, education, and economic opportunity. ASFT is founded on the belief that a community's knowledge of, economic dependence on, and respect for fisheries resources engenders a strong sense of stewardship and provides a vital voice for conservation. Maintaining and securing fishing permits or quota in communities can create mechanisms for incentivizing sustainable or conservation-focused fishing practices, such as sharing information to avoid bycatch or protect sensitive habitats. Securing local access can also preserve fishing infrastructure, help maintain a diverse and active fleet, provide opportunities for future generations, and promote economic stability.

ASFT's Local Fish Fund LLC (LFF) provides bridge financing for intergenerational quota share (QS) and Alaska limited entry permit transfers to recover and retain fisheries access for Alaskan rural residents. LFF connects willing sellers of QS and limited entry permits with Alaskan fishermen buyers who would not otherwise qualify for traditional financing. LFF allows fishermen, community residents, or others who support the mission of the ASFT to participate in loans to community-based fishermen, sharing the high costs of entry and the risks faced by new fishermen. To advance sustainable fishing practices and community objectives, LFF fishermen are also required to participate in initiatives designed to contribute in a positive and tangible way to sustainable fisheries management.

ASFT recognizes and supports the importance of anchoring access in communities through community development quotas (CDQs), community quota entities (CQEs), and permit banks. ASFT is also committed to supporting Alaska's independent fishermen as an integral part of Alaska's fishing future. We believe coastal fishermen are important community leaders.

ASFT recognizes:

- Fishermen are independent people who want to control their future.
- Successful fishing businesses need equity to grow.
- Independent fishermen are advocates for healthy fisheries and healthy fishing communities.
- QS investment deepens commitment to resource stewardship.

ASFT was formed to promote and support access for communities **and** independent fishermen. In forming ASFT, we worked to overcome the barriers independent fishermen face when purchasing QS.

In a recent study, halibut fishermen reported their reasons for not purchasing halibut quota share and listed concerns about charter catches, difficulty with obtaining financing, and concerns about annual limits as primary obstacles to entry.

Barriers to QS Purchase

Table 1.16 Plans for not purchasing more halibut QS.

	No opinion	Strongly disagree	Somewhat disagree	Not a factor	Somewhat agree	Strongly agree	Total responses
Cannot afford more QS at this time	0.04	0.05	0.05	0.22	0.3	0.33	276
Not enough time	0.04	0.35	0.06	0.43	0.09	0.03	275
Planning to Retire in the near future	0.06	0.26	0.03	0.38	0.19	0.08	279
Concerned about annual limits	0.03	0.01	0.02	0.09	0.35	0.5	282
Concerned about charter catches	0.02	0.03	0	0.08	0.19	0.68	281
Too difficult to obtain financing	0.05	0.08	0.05	0.43	0.25	0.12	276
Plan to buy QS in a different fishery	0.19	0.19	0.04	0.43	0.11	0.03	272

Source: Kotlarov. A. 2015 Characterizing Crew and Fuel Price Impacts: A Survey of Pacific Halibut and Sablefish Quota. Share Holders, Technical Memorandums. NMFS-F/AKR 11, 89 p. https://alaskafisheries.noaa.gov/sites/default/files/reports/noaa-tm-akr11-sablefish-halibut-qsholders.pdf

Standard financing is scary to new fishermen and community quota entities because the down payment requirement can be staggering (20% or higher), and payments are fixed regardless of fluctuations in fish abundance, fish prices, or QS value.

The table on the following page provides an example of why standard financing is a risky business for new fishermen. The numbers are typical of the cash flow for a fisherman who recently purchased sablefish quota in the Southeast area and factors in standard costs, QS payment, crew share, etc. The example establishes that on a $100,000 investment the take home pay is just $1,762. Of course, the fisherman is building some equity in the shares and supporting some vessel costs, but the paycheck after fishing the shares quickly goes in the red if costs go up or the quota goes down. For good reason, new entrants are hesitant to take the risks.

To address these concerns, ASFT launched Local Fish Fund, LLC. The Local Fish Fund is a fishery access participation pool. LFF provides bridge financing for intergenerational quota share transfers to restore access opportunities to fishermen in Alaska's coastal communities. LFF connects QS sellers with community-based fishermen and fishermen

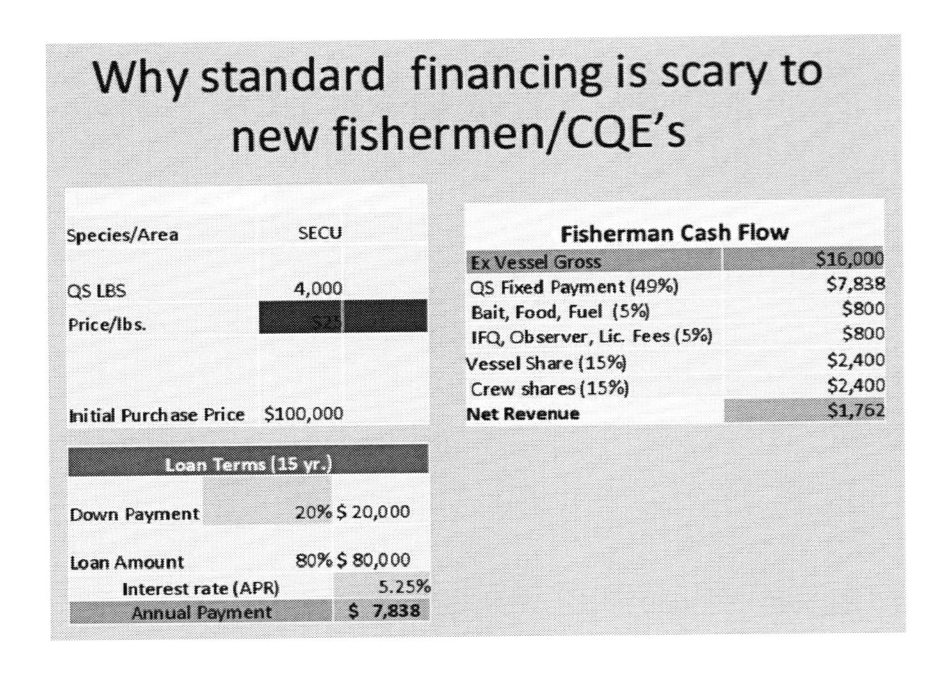

or foundations willing to share the risk with new entrants. LFF helps young fishermen buy QS and generate the equity to refinance through a traditional lender, while sharing the risk from QS price variance. In return, LFF receives a portion of fishing revenue and QS appreciation. The buyer's risk is capped, while the initial capital of those who participate in the QS purchase is shielded.

LFF is working with Haa Aaní and The Nature Conservancy to explore opportunities to collaboratively participate in the future of Alaska sustainable fisheries and sustainable fishing communities.

Local Fish Fund goals are to (1) provide a clear path to ownership for independent, entry level fishermen; (2) provide competitive loan terms; (3) overcome barriers through shared risk and shared gains; and (4) provide modest returns while minimizing risk of capital loss.

How does LFF work? The buyer pays 10% of the purchasing price as a security retainer. The QS seller is fully compensated for the QS value by those participating in the loan. Loan participants share in annual fishing revenue generated by QS (50% to 60% of the ex-vessel, or dock price paid to fishermen when fish are delivered). At loan term the buyer refinances or sells QS. By this time the buyer has a successful fishing and borrowing history. The loan is structured to result in 20% buyer equity under most circumstances. The buyer and loan participants both share the QS value gain.

Why does LFF work? The LFF process meets the needs of the QS sellers, by providing market price, and meets the needs of QS or limited entry permit buyers by reducing entry level costs and lowering the risks faced by buyers.

For QS buyers, LFF reduces upfront costs and the risks associated with fixed payments. The buyer pays 10% of purchasing price as a security retainer. LFF provides a clear path to ownership. At the loan term, the goal for the QS buyer is 20% equity in shares, plus fishing and borrowing history. LFF manages QS buyer risk. Payments are based on ex-vessel revenue, not fixed amounts; risk is limited to 10% security retainer; and the commitment/exposure time is reduced to five years.

LFF also manages loan participants' risk. The fishing revenue and QS value ensure modest return to participants across a broad range of QS and ex-vessel price changes, and investors share QS upside gains.

Where does the LFF model work? LFF works for independent fishermen purchasing QS, for CQEs purchasing QS, and for independent fishermen purchasing state limited entry permits. LFF has partnered with Haa Aaní to facilitate limited entry permit transfer, and we hope that LFF can partner with the State of Alaska or Commercial Fishing and Agriculture Bank (CFAB) to finance permit purchase.

ASFT vision recap

- The Local Fish Fund provides independent fishermen with a clear path to QS ownership.
- Through LFF, ASFT helps small boat fishermen secure a voice in fisheries management.
- Independent fishermen are important community leaders.
- ASFT supports fisherman engagement in resource management and conservation.
- By engaging community-based fishermen in fisheries' management, ASFT will promote policy scaled to meet the needs of small boat fishermen.

In closing, the factors driving loss of fishery access from coastal communities are complex. We hope ASFT's Local Fish Fund can be part of the solution that reverses this trend.

Kelly Wachowicz and Eric Hesse address the group. (Deborah Mercy photo.)

Establishing a Community Fishing Association in the Developing Gulf of Alaska Trawl Bycatch Management Program

Theresa Peterson
Alaska Marine Conservation Council, Kodiak, Alaska

Ernie Weiss
Aleutians East Borough, Anchorage, Alaska

Community fishing associations is an idea that presents a solution to the barriers to entry so often found in catch share or limited access privilege programs.

Gulf of Alaska communities

Gulf of Alaska communities are all very unique, but they are similar in their need to access fisheries resources outside their front doors to maintain a stable fishing economy. The Gulf of Alaska is not the Bering Sea, and fisheries management needs to be designed with the long term health of our communities front and center.

Kodiak is a magnificent Island that lies in the middle of some of the richest fishing grounds in the world. The island is home to about 13,500 people, living in the city of Kodiak and in six villages on the islands. Kodiak has the largest and most diversified port in Alaska. The major fisheries are crab, herring, halibut, groundfish, and salmon. Kodiak ranks among the top ten fishing ports in the United States for commercial fishery landings and value, and one out of every three jobs in Kodiak is directly connected to fishing industry.

Fishing has provided food and sustenance to the people of Kodiak for thousands of years, and in more recent history commercial fishing has become the backbone of our economy. As islanders we are proud of our community and have a deep desire to see the legacy of subsistence and commercial fishing continue for generations to come.

The Aleutians East Borough (AEB) encompasses six coastal communities: Akutan, False Pass, Nelson Lagoon, Cold Bay, Sand Point, and King Cove. King Cove is on the Alaska Peninsula, and Sand Point is on Popof Island. Sand Point has an estimated 21 active groundfish LLPs (License Limitation Program), and King Cove has 15 active LLPs. The trawl communities are Sand Point, population 976, and King Cove, population 938. There are a total of 18 trawl catcher vessels between these two communities. In contrast, False Pass is a small community with a population of 35. It is in the Gulf of Alaska but is considered a Bering Sea CDQ (Community Development Quota) community, and it has no trawl vessels. All three of these communities are served by the Alaska Marine Highway System, but False Pass receives about half the port calls that the larger communities see. False Pass has just one boat harbor compared to two harbors each in Sand Point and King Cove. King Cove has a 3,500 foot gravel runway and Sand Point has a 5,200 foot asphalt runway. All three communities have only one seafood processor each—Peter Pan in King Cove and Trident in Sand Point are both year-round facilities, while Bering Pacific in False Pass processes much less seafood. The August 2015 social impact assessment for the Western Gulf of Alaska trawl bycatch management can be found at www.aebfish.org.

Fisheries rationalization to address bycatch

We want to share the discussion occurring at the North Pacific Fishery Management Council and why we support a community fishing association as part of the discussion. In 2013 the Council began the process to develop a new program for Gulf of Alaska trawl fisheries. The following is from the Purpose and Need Statement: "Management of Gulf of Alaska (GOA) groundfish trawl fisheries has grown increasingly complicated in recent years due to the implementation of measures to protect Steller sea lions and reduced Pacific halibut and Chinook salmon Prohibited Species Catch (PSC).... The new program shall be designed to provide tools for the effective management and reduction of PSC and bycatch, and promote increased utilization of both target and secondary species harvested in the GOA."

The purpose and need statement from the Council clearly outlines the desire to provide a new management structure in Gulf of Alaska trawl fisheries due to increased challenges facing the trawl fleet and the need to reduce bycatch of valuable species of salmon, crab, and halibut. The species caught as bycatch are important to commercial, charter, recreational, and subsistence fishermen in communities throughout the Gulf of Alaska. A new management system could also increase utilization of both target and secondary species harvested and processed.

While there is a common desire to meet the purpose and need, how you get there is another story.

The Council's attempt at rationalizing the Gulf of Alaska trawl fishery was to address significant bycatch issues, where catch shares are promoted as a bycatch "tool." A catch share program allocates a specific portion of the fishery resource to participants. Catch shares are a tool for promoting economic efficiency and they can also serve as a tool for bycatch reduction by allowing the fleet to slow fishing down (i.e., effectively ending the "race for fish") and make better decisions about when and where to fish relevant to bycatch.

But catch shares are controversial due to negative impacts to communities and working fishermen:

- Absentee ownership of quota.
- High quota leasing fees.
- Consolidation of vessels and quota ownership.
- Lower crew pay and job loss.
- Out-migration of fishing rights and wealth from fishery-dependent communities.

Attempts at implementing catch shares over the last decade have failed due in part to the contentious aspects of privatizing a public resource.

When Governor Walker took office in 2014, he appointed Sam Cotten as Commissioner of Fish and Game. In addition to many other responsibilities, Cotten serves on the Council. After requesting a delay of further development while the new administration reviewed the program, Commissioner Cotten broadened the alternatives for development and added an alternative that allocates only prohibited species of halibut and Chinook salmon to cooperative participants. This approach would provide individual vessel accountability within the cooperative structure without allocating the target fishing rights. The new alternative is under analysis and we are excited to have an alternative to consider that provides the tools without privatizing the fishery.

Direct allocation to communities

In October 2014 the Council added a community fishing association alternative to analyze within the proposed catch share alternative. The Council alternative outlines the fundamental elements for consideration in developing the community fishing association, such as initial allocation percentages, goals, and objective and other critical information needed. However, the framework requires a substantial amount of work to further define the concept.

The alternative outlines elements including:

- Proposed CFA allocation range of 5-15% for analysis.
- Goals and objectives.
- Community eligibility considerations.
- Elements to include in a community sustainability plan.
- Requirements for an annual report.
- Description of how the CFA will integrate with the coop program.

The Council goals and objectives are:

1. Authorize fair and equitable access privileges that take into consideration the value of assets and investments in the fishery and dependency on the fishery for harvesters, processors, and communities.
2. Balance interests of all sectors and provide equitable distribution of benefits and similar opportunities for increased value.
3. Promote community stability and minimize adverse economic impacts by limiting consolidation, providing employment and entry opportunities, and increasing the economic viability of groundfish harvesters, processors, and support industries.
4. Promote active participation by owners of harvest vessels and fishing privileges.

Direct allocation to a community fishing association would allow for the benefits of a quota program to be realized (i.e., slowing down the race for fish), but would mitigate some of the negative social and economic consequences of catch share programs by limiting the commodification and transferability of quota. Community held quota could provide affordable access opportunity for fishermen trying to enter the fishery without the capital to buy quota. Community held quota could be structured to address unintended consequences that arise after a program has been implemented.

Community fishing association concept and structure

The central tenet of the community fishing association concept is direct allocation of quota to an association in order to anchor quota in communities in perpetuity. Direct allocation to fishing communities was authorized in the 2006 reauthorization of the Magnuson-Stevens Act (MSA) §303a(c)(3) under "fishing communities" in the limited access privileges section. In addition to meeting fishing community eligibility, communities interested in holding quota must submit a community sustainability plan that demonstrates how the plan will address the social and economic development needs of coastal communities.

Key and unique benefits of a direct allocation to communities:

1. Ensures the long-term multiplier of benefits, especially fish dollars, to local communities through the anchoring of quota to communities in perpetuity.

2. Acknowledges crew as real stakeholders (e.g., through crew contracts and codes of conduct to protect crew from reduced compensation).

3. Supports entry of fishermen without adequate capital.

4. Includes fishermen and community perspectives in operation and decision-making processes that impact local communities and businesses (i.e., ensures that local communities benefit from local fishery resources).

Implementing community fishing association allocation

The Aleutians East Borough applied for and received a grant from the National Fish and Wildlife Foundation that will assist stakeholders in establishing a community fishing association in the Gulf of Alaska to provide long-term fishing community access and stability. The grant will aid in setting up a steering committee and a community workgroup to guide the outline of a community sustainability plan, including the governance structure. Grant funding may also support establishing the community fishing association as a legal entity able to receive and hold quota.

Conclusion

From the outset, community fishing association proponents have sought a structure that offers benefits and opportunities for fishing communities beyond the community stability measures presented in the Council's current program design. The main community protection measures in the Council's program design are: (1) consolidation limits, (2) regionalization of landings, and (3) active participation requirements. Each of these represents a critical source of community stability, but none sufficiently address or mitigate the impacts associated with changes such as the flight of quota from communities or equitable crew compensation.

One vision of a community fishing association in the North Pacific Fishery Management Council's proposal for a Gulf of Alaska trawl bycatch program is to reserve future opportunity for communities like

False Pass that haven't yet been able to participate in the trawl fishery. Another view is that the community fishing association is meant to protect those who have invested and are now dependent on the trawl fishery. In the end, we are looking for solutions and the community fishing association represents one idea, if federal policy makers decide a catch share is the preferred means to address bycatch in the Gulf of Alaska.

Maine's Lobster Licensing Program

Deirdre Gilbert
Maine Department of Marine Resources, Augusta, Maine

Maine's coastal economy and identity are closely tied to the lobster industry. Lobster makes up 78% of the commercial fisheries landings, with an ex-vessel value of $458,372,699 in 2014.

The Maine lobster industry today is conducted by owner-operators, mostly in day-boat operations. There are about 5,000 license holders. It is a year-round fishery—about 80% of landings are between July and November.

The economic value of the lobster fishery, and pounds harvested, have seen a sharp increase since 1990. The number of lobster licenses

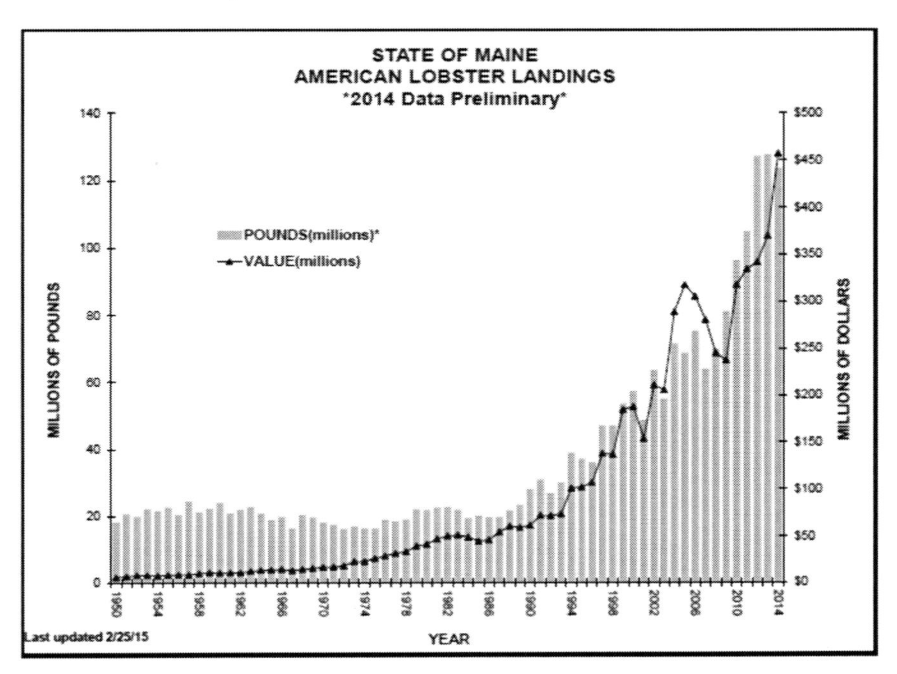

has decreased from 6,400 in 1997, when limited entry was established, to 4,900 in 2014.

Limited entry for lobster

In 1997 all lobster license holders had to qualify for a license. After 1997 they either had to have held a license in the prior calendar year, or qualify through the student or apprenticeship programs. Zones may be "open" or "closed." For closed zones, entry is based on an exit to entry ratio of 3:1 or 5:1.

Student program

The student program was created at the same time as Limited Entry, because the industry wanted to preserve the traditional method of access for young people growing up on the coast of Maine. A qualifying individual must be a full-time student. Students must complete the Apprenticeship Program and purchase their license before they turn 18 to avoid being on the waiting list. Students can fish up to 150 traps.

Apprenticeship program

To complete the Apprenticeship Program a student must fish 200 days and/or 1,000 hours over a minimum of two years. Drill conductor safety training is mandatory. An apprentice can have up to three sponsors. A sponsor must have held a license for minimum of five years. Apprentices must work in the zone they wish to enter. An apprentice must have their logs signed by a sponsor and local Marine Patrol Officer, and submit them every 250 hours, within 30 days of last day logged. If apprentices are over the age of 18 years when they complete the program, they must go on a waiting list

Island Limited Entry Program

There are 13 Maine islands with year-round communities; five Islands are currently in the Island Limited Entry Program and one is in the process of exiting. Island residency is a requirement. Island lobster license holders may vote for their island to be part of the Island Limited Entry Program, and a two-thirds vote is required. The number of new island licenses each year uses a 1:1 entry to exit ratio. Fishermen must live and fish from the island for a minimum of eight years before being able to take the license off the island.

Iceland's Experience: Community Quota and Coastal Fishing

Catherine Chambers
Hólar University College, Blönduós, Iceland, and University of Alaska Fairbanks

Introduction

This talk reviews two programs in Iceland, the community quota program and the coastal fishing season, in terms of their ability to provide access to fisheries resources. The majority of Icelandic fisheries have been under a nationwide ITQ (individual transferable quota) system since the 1980s, and problems with consolidation of fishing rights and rural decline continue to exist. There is also a limited license lumpfish fishery that is outside the ITQ system and the new option of the "coastal fishing" season that is reviewed here.

ITQs are split into a large-boat system and a small-boat system. In 2015, there were 267 large boats (defined as over 30 GT [gross tonnage]) that caught 92% of the total catch landed in Iceland. Small boats number around 1,400, and account for the remaining 8% of the total catch; less than 1% of the total catch is from the non-ITQ fisheries of lumpfish and coastal fishing. The governance system is relatively streamlined, where all fisheries fall under the Ministry of Industry and Innovation that takes official scientific advice from the Marine Research Institute.

Because small-boat fisheries and small rural communities experienced more negative impacts from the ITQ system, this talk focuses on small-scale fisheries. In terms of access, my research has found small-boat fisheries are considered an important part of Icelandic national identity and very much ingrained in family cultural history and identity. On a mailed survey I sent to small-boat fishermen all over Iceland, I found that the average small-boat fisherman has three generations of family not including himself involved in fishing, but many just wrote in answers like "since the oldest men remember." At the same time, there is an ongoing "graying of the fleet" phenomenon. The survey showed that less than 1% were under 30 years old or had less than five years of experience. Current small-boat fishermen are on average 58 years old and have over 30 years of fishing experience. Around half of them, however, worked as crew at some time in their past, averaging around 13 years of crew experience. This means they are the last generation of small-boat fishermen to work their way up through the traditional path of a fishing career—from crew to skipper to owner.

Community quota

Community quota was officially enacted in 2003. It has the major goal of bringing economic benefit to coastal communities that were negatively affected by the ITQ program. Each year, the Ministry gives a small percent of the total quota directly to boats which then must land that quota in their home port. In 2015, less than 2% of the cod TAC (total allowable catch) was given to the community quota program. The fishermen send in an application directly to the Directorate of Fisheries for the quota, but the municipalities themselves can also request special rules for the recipients of the quota. For example, one municipality made the rule that boats which lost their shrimp quota when the fishery collapsed in 2001 should each receive a certain amount of community quota to support them since the loss of the shrimp fishing was a big blow to their businesses. Each municipality can request rules that they see as best benefiting their community: gear restrictions, boat size restrictions, species restrictions, landing at particular processors, and so on.

The community quota used to be given directly to the municipality, but this process was extremely politically charged because the community leaders were tasked with distributing the quota in the way that best benefited the community as a whole. This means that they often gave quota to individuals they knew would fish the most valuable fish, bringing in the most taxable revenue for the community. The way it is set up now surpasses that difficult political process, but at the same time it means it is basically just bonus quota given to those who already fish. It is not a community ownership or pool that could be used as collateral to build up for other fisheries, or to support newcomers. And finally, the municipality still makes the rules, so the fishermen don't really have an official say in what happens with the community quota.

Coastal fishing

The coastal fishing season began in 2009, with double goals: (1) to open up access to fisheries as a response to a 2004 UN Human Rights Committee ruling that said the ITQ system violated the human right to work; and (2) to offer economic development opportunities in the rural fishing villages. Fishing takes place from May to August, Monday through Thursday, with a 14-hour time limit. Four regions around the country each have their own monthly catch limit. Fishers can bring in up to 650 kg of bottom fish per day, using no more than four jig machines. When the region catch limit is reached, the entire area closes down until the beginning of the next month. This fishing time gets smaller as more boats take part: in August 2015, for example, the limit of each of the four areas was caught in seven days.

Boats are typically run by a single person. The numbers of participating boats fluctuates around 550-750 for the whole country. There is a rule that a fisher cannot fish ITQ fisheries and coastal fishing at the same time, but a person can have quota, finish it for the season, and then switch into coastal fishing. In 2012, for example, 460 of the 664 coastal fishing boats fished only coastal fishing, but 114 combined coastal fishing and lumpfish, and 90 did have quota.

I carried out some research to determine how current fishermen were thinking about who coastal fishing **should** be for because there is a lot of confusion about the actual purpose of the system. I came up with four major categories that explain the differences that fishermen and communities express in their perceptions of the coastal fishing season. Some thought coastal fishing should only be to live the "Icelandic Dream": an older person fishing because he just needed to be at sea, or a way for former fishermen to retire while still fishing a little. Others thought coastal fishing was full of "American Dream" fishermen: people in the system to make money and who already had other jobs. Still others thought the system could be either for ITQ holders whose quota was so small that this system could help them, or for non-ITQ holders who didn't have enough money to purchase quota.

My survey found that the average coastal fisher is 60 years old, with 30 years of fishing experience and 25% of their yearly income from fisheries, so it seems as though whatever the intention of the system was, the "Icelandic Dream" or perhaps "American Dream" fishers are prevalent in the fishery.

Another way I could analyze the coastal fishing season was to ask the question "Would you advise a young person to enter your fishery?" Coastal fisheries participants answered "no" more often than quota holders or lumpfish fishermen, for reasons such as a high startup cost, politics and stress, low salary, and that it was "not worth the effort." Similarly, I asked how fishermen agreed with the statement "I have the flexibility to join other fisheries if my primary fishery is not doing well in any given season." Coastal fishers disagreed most with this statement, suggesting coastal fishing is not good for newcomers intent on building up a fishing business.

Conclusion

So you could say that coastal fishing seems to most benefit individuals who are already engaged in fishing, or who are financially established in other trades or professions. This means in terms of access to newcomers the system is not performing well. It has a high cost, it is inflexible, and because it is a single-person fishery there is not the chance for newcomers to learn fishing or business skills. However, when viewed from the community itself, the coastal fishing season is a great positive benefit. The community can collect harbor fees from the extra boats, fuel and supplies are purchased in town, and there is general activity at the harbor that community members like to see. This is somewhat similar to the community quota—there are economic benefits to the community and to certain individuals, but the question with these kinds of programs is **who** they should be designed to benefit. Is it the economic situation of the community, is it any fishing business, is it newcomers or youth? How do we measure success, and is there a way we can benefit the community economy as well as provide access to newcomers at the same time?

For more information:

Catherine Chambers and Courtney Carothers. 2016. Thirty years after privatization: A survey of Icelandic small-boat fishermen. Marine Policy. http://dx.doi.org/10.1016/j.marpol.2016.02.026

A Milder Version of ITQs? Post-ITQ Provisions in Norway's Fisheries

Einar Eythórsson
Norwegian Institute for Cultural Heritage Research, Tromsø, Norway

This presentation is a brief outline of the provisions that have been in place in Norwegian fisheries since the introduction of individual vessel quotas for the coastal fleet in 1990. It is contested whether the Norwegian quota system is an individual transferable quota (ITQ) system; transferability of quota is restricted in Norway and the fisheries authorities avoid using the ITQ label. The long-term robustness of limitations to transferability is also addressed in the presentation.

ITQs in Norway

In Norway vessel quotas (IVQs) have been in effect for the **offshore fleet** since the 1980s, and for the **coastal fleet** since 1990. The initial allocation was according to vessel length, while requiring certain historical catches. IVQs were non-transferable at first, and since then there has been a gradual development toward (limited) transferability. There are separate schemes for offshore vessels (trawlers), and inshore/coastal vessels. Inshore total allowable catch (TAC) (about 70% of the total) is distributed to a **closed IVQ group** and an **"open" group** without IVQs. The closed IVQ group is divided into five subgroups defined according to vessel length.

Transfer limitations for the IVQ groups

IVQs cannot be transferred as a separate commodity—rather they must follow a vessel. IVQs cannot be transferred between counties unless the vessel owner has moved to another county. IVQs are transferable within **subgroups,** but not between them. Quota leasing is not allowed. Quotas are transferred for a limited term only (10-18 years). IVQs can only be transferred to another vessel if the first vessel is permanently removed from the fishery. Only 80% can be transferred; 20% is then redistributed within the subgroup. No transfer is allowed in the subgroup for vessels under 11 m in length. A buyout program has been carried out for this group, co-financed by the government and the fisheries.

The guiding idea behind the policy of limited transferability is to slow down the concentration of IVQs, geographically and within large companies, and to preserve a diverse fleet structure. Another impact of limited transferability is to slow down an increase in IVQ prices.

These limitations have been quite effective in terms of preserving a diverse coastal fleet. For the offshore fleet, however, a huge concentration of ownership has taken place, through company mergers and takeovers rather than direct quota transfers.

There may be a reason to ask how robust the limitations to quota transfers will be in the time to come. Loopholes exist, and in the long run tension builds up within the fisheries. There are increasing complaints about the complexity and inflexibility of the current system. Quota assets have acquired a market value that can be realized through sales and used as collateral for loans. For each individual quota holder, fewer restrictions would be beneficial in the form of higher market value of their assets. For fishers as a group, and especially for fisheries communities, liberalization of transferability represents a risk and a loss of predictability about their future livelihood. The transferability restrictions will probably survive only as long they are supported by the fishers' organizations.

District quotas

Fishing quotas in Norway are generally not attached to a region or a community. In 2006-2007 there was an experimental distribution of **"district quota"** in northern Norway. IVQs in three subgroups: offshore, 21-28 m vessels, and 15-21 m vessels were cut by 3%. This quota was redistributed within the same subgroups, according to recommendations from municipal and county councils. The idea was to allow for more participation in quota allocation by regional and local authorities, and to address local issues. The arrangement was not evaluated as successful, and was discontinued after 2007.

Recruitment quotas

The purpose of recruitment quotas is to recruit young boat owners. This quota system applies to the 11-15 m vessel length subgroup. About 10-20 individual fishers have received recruitment quotas free of charge during 2010-2015. Recruitment quotas are "funded" by redistribution within this subgroup.

Recreational fishing, tourist fishing, and youth fishing

Recreational fishing and youth fishing are allowed within gear restrictions. For tourist fishing, an upper limit is applied for how much fish can be exported by foreign tourists. Youth (age 12-25) can apply for permission to fish during summer holidays (June 21–August 13) with hand-line, longline, or gillnets. Each year 3,000 tons of cod from the annual TAC is set aside to cover these three categories. There are no reliable statistics available on catches from these fisheries.

The open group fishery

The open group was initially a compromise to accommodate part-time and small-scale fishers who did not qualify for IVQs in 1990. Fishers who own a small fishing vessel and have less than 4G (about $40,000 USD) annual income from other (non-fishery) sources are qualified to apply for permission to fish in the open group. Maximum annual catch (according to boat length) is not guaranteed—it is an "olympic" fishery for a certain fraction of the TAC. The open group has worked as a viable alternative for small-scale fishers during the last decade; despite being open to new entrants, it is not yet overcrowded.

Indigenous population special provisions

There are special provisions for districts with indigenous (Saami) populations. Since 2011, **3,000 tons of cod** have been set aside annually to increase the maximum catch for small-scale, open group fishers in Finnmark county and parts of Troms and Nordland counties. For 2016, this means that open group fishers within Saami districts can catch a quantity of cod equal to IVQ fishers in the under 11 m vessel length subgroup.

Alaska king crab

Alaska king crab were introduced to the Atlantic Ocean by Russian scientists in the 1960s. A commercial fishery for the crabs has been in effect in Norway since 2002, in east Finnmark only. Since 2008, the crab quota has been distributed annually to all applying fishers in east Finnmark, including small-scale open group fishers.

How robust are these provisions?

All special allocations from the total allowable catch are unpopular among IVQ fishers. TAC for cod has been high and has increased during the last decade, which probably has increased the tolerance for special TAC allocations. The open group has become part of a compromise between the Saami Parliament and the Norway government; this increases its chance to survive.

The Norwegian IVQ regime can be seen as a "light" version of ITQs. The restricted transferability and the different provisions are the result of a series of compromises, and some of them are likely to disappear in the long run. The open group fishery and the special provisions for Saami districts may survive because of the international attention toward Norwegian practices regarding indigenous access to the fishery.

ITQ in Greenlandic Fisheries: Access Issues Addressed/ Unaddressed?

Rikke Becker Jacobsen
Aalborg University, Denmark

Greenlandic prawn fishery

The Greenlandic prawn fishery has been managed by individual transferable quotas (ITQs) since 1990 and 1996 for the offshore and coastal segment respectively. From 1991 to 2011 the prawn fleet was reduced from 42 trawlers to 5 highly efficient and capital-intensive factory trawlers. Concerns over the concentration re-emerge in the Greenlandic Parliament from time to time, but no measures have been taken to counter the development, and ITQ as a management tool still receives support from influential groups in Greenlandic society. Thus, in 2012 ITQ management was introduced into parts of the coastal Greenland halibut fishery with the aim of increasing profitability in the fishery. Addressing issues of future community access to fishing quotas that may follow from the ITQ reform are only marginally raised and discussed, e.g., by looking into possibilities of small-scale fishers teaming up to buy and manage quota and being able to finance larger fishing vessels.

Greenlandic halibut fishery

The Greenlandic halibut fishery has more than 1,000 small-scale participants representing small, spread-out settlements in North-east Greenland. In 2012 the three quota districts were divided into dinghies in "olympic" competition and cutters with ITQ.

Long-term access issues in ITQ are not addressed. The focus is the national economy (self-governance) and the size of taxable incomes in the fishery (bigger-is-better rationale). Addressing access issues has not been a priority. In general there is a prevailing optimism that other employment and life opportunities would materialize for the population in the long term. In 2009-2011 there were high hopes for oil and mining.

Even today the long-term access ITQ issues are only marginally raised as a problem. When the issues are raised, the main interest is in seeing small-scale fishers team up, such as creating cooperatives.

When short-term access issues are addressed through the olympic fishery, the dinghy fishers experience quota and access restrictions immediately, and they have very strong political power. Short-term access issues are "solved" by new politicians/government who annul closure of new access, raise the olympic quota, and establish new free-from-TAC (total allowable catch) fishing areas.

Permit Banking on Cape Cod

Eric Hesse
Cape Cod Fishermen's Alliance, West Barnstable, Massachusetts

Cape Cod Fisheries Trust

I am a board member and past president of the Cape Cod Fishermen's Alliance. The Cape Cod Fisheries Trust is a program of the CCFA. This contribution is about my experience both in setting up the trust as a founding board member and in using the trust as a fisherman. The driving force behind CCFT was Paul Parker, who continues to manage it. CCFA formed in 1991 to advocate on behalf of Cape Cod's small-boat hook fleet.

The New England multispecies fishery for cod and haddock became limited entry for all vessels in 1994 but was managed under daily trip limits and limits on days at sea until 2010. The Cape Cod Fishermen's Alliance formed the Georges Bank Hook Sector as an experimental quota-driven fishery. The sector was allocated quota based on a ten year span of its members' fishing history. An individual's history, known as a Potential Sector Contribution or PSC, is only valuable under the sector management umbrella. "Common pool" vessels still fish under days at sea and trip limits. The trust, and now the Nature Conservancy, hold permits managed by the Georges Bank fixed gear sector.

My fishing background

My two boats are a vestige of the days of open access and trip limits, which led to overcapitalization. Most small-boat New England fishermen use downeast lobster-boat designs, house forward with a deep keel and planing capability. I use a 40x16 foot Wayne Beal boat for hook fishing and research work and a 30 foot South Shore (designed by Calvin Beal) for harpooning tuna.

I started fishing 32 years ago. I crewed and ran others' boats for seven years before building in 1991. I have fished mostly in the Gulf of Maine and Georges Bank. I love to fish and have built my life around it.

Fisheries management

A brief overview of our fishery areas is shown on the map. This is just one layer of the lines drawn in the ocean for various reasons. An important line is the one that extends east from the northern tip of Cape Cod to the Hague Line. It is the dividing line between Georges Bank and Gulf of Maine fishery areas. Our historic winter cod fishery off Nantucket collapsed in 2002, forcing smaller boats offshore to Georges Bank.

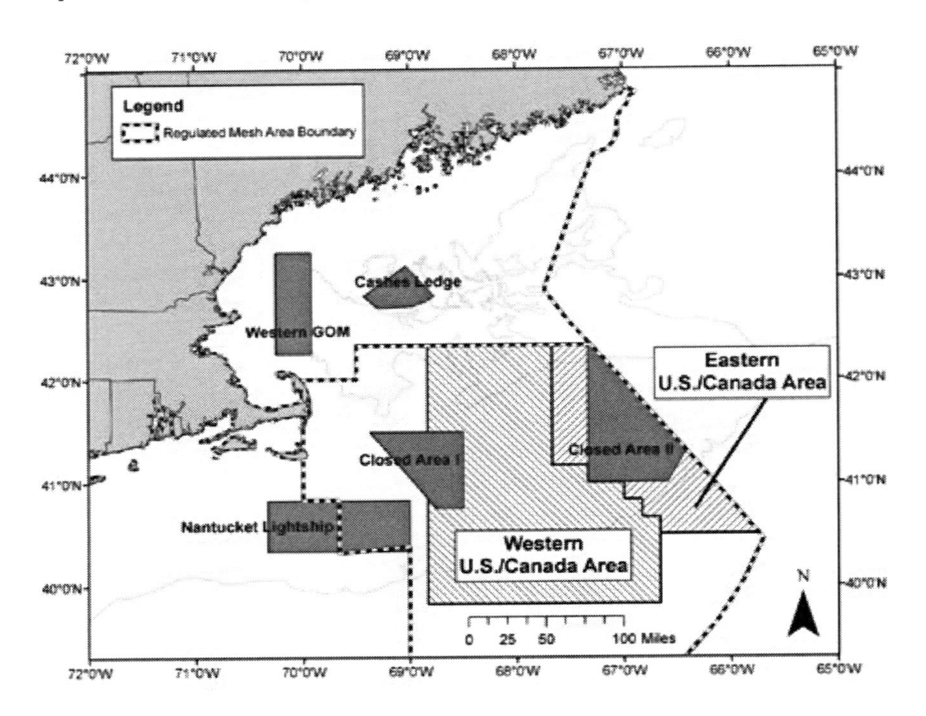

Diversified fleet

Our fleet is diversified to exploit migratory, seasonally available inshore fish populations. Tuna, groundfish, and dogfish are the mainstays. A typical mix of fisheries for local boats also includes striped bass, skate and monkfish (gillnet), bluefish, mackerel, and of course lobster.

Our tuna fishery is nothing like what people may see on *Wicked Tuna*—there is much less drama. The average price last season was $4.50 per lb. I have used a spotter plane some years and not others. There has been a general decrease in average fish size; now they are bigger and more plentiful off eastern Nova Scotia.

Dogfish are "trash fish" exported to Europe. We are working hard to develop domestic markets, constrained by processor and Marine

Stewardship Council bottlenecks. The ex-vessel price averaged $0.20 per pound last season.

Research work

Research helps fill the void left by the winter cod fishery and broadens our understanding of local resources. We tag halibut, cod, tuna, ocean sunfish, haddock (for bycatch research), basking sharks, great white sharks, and sea turtles using satellite and traditional streamers. Cod genetics studies look for markers associated with distinct subpopulations that may exhibit spawning site fidelity.

Longline stock assessment research supplements the traditional trawl survey, and is better for some species and bottom types. The East Coast halibut fishery is poised for a comeback but data are lacking. Thorny skate is currently being investigated for ESL (environmentally sensitive localities) status, and cusk is a subject of study because it is a data-poor species.

Setting up the Cape Cod Fisheries Trust

My involvement in setting up CCFT could easily be compared to making sausage. It was never pretty! From the beginning Paul Parker encouraged the board to take a long view, to look past our own noses, and to stay true to our core values and mission. We trusted Paul as a member of our fishing community and a very sharp guy.

The trust currently serves about 50 vessels and 120 fishermen and crew in the Cape Cod fishing community. The trust has surveyed the fishing community multiple times to learn what members needed to plan their business and move forward.

Committing to obtaining the capital required to purchase permits was a particular leap of faith for my fellow fishermen. It helped to have foundations and later donors who liked and understood our story. The experience foundations have had with CCFT helped pave the way for similar investment in other communities.

Our local groundfish fishery was collapsing as the lease model rolled out. It was fortuitous in a way since as a result there was less demand and things were less contentious. The model was perfected and then plugged in to scallops, which is a healthy fishery now.

The actual mechanics of setting up the trust are discussed in the publication *Banking on Your Fishing Future*.

Trust milestones

Trust milestones involved strategic planning, initial allocations, and investment policy. Milestones that stand out for me include obtaining capital: Ford, Calvert, Cape Cod, and Campbell all took a big risk on us from my perspective as a board member. We had a good story and were part of the fabric of the Cape's community. Another milestone was moving from passive to active portfolio management.

As cod resource declined we realized a diversified portfolio would be safer and more reliable and match our fishery goals. We moved in to scallop, surf clam, and trawl-caught species that we had interacted with infrequently. Further stages involved leasing quota inside and outside sector, and development of investment policy with species overall allocation caps. The community had lots of Georges Bank cod it didn't need or couldn't catch, and not enough of other things. For tasks such as balancing the portfolio for the long term, you never get it right the first time! We often took three steps back before moving ahead.

Scallop is an interesting example. It is a hot species in New England now. We are keeping our waterfront infrastructure alive and providing hope for the future. It is a complex system to decide who gets what of limited scallop quota. Lease prices were set at 50% of market for sector vessels, which pays interest and more! It is a very successful program with high demand.

Using quota from groundfish

How did we allocate trust resources to sector fishermen? We aimed to avoid conflict among fishermen over allocation. In the waning days of the cod fishery we found several good hook fishing opportunities to catch cod and haddock. One of these was when Amendment 16 to the Northeast Multispecies Fishery Management Plan was approved. Vessels in our Georges Bank sector were allowed to fish in the Gulf of Maine for the first time in six years. We had almost zero Gulf of Maine allocation at that time.

Before we headed to the Gulf of Maine we talked with trust manager Paul Parker regarding our estimate of quota needs. A typical trip for us can run anywhere from 2,000 to 5,000 pounds per day of cod and haddock. He was able to lease us 20,000 pounds of Gulf of Maine cod from the northern sectors, and two boats headed north to Stellwagen. We fished all day for 50 pounds of cod on the jig, never finding a spot that looked good enough to set our 3,000 baited hooks. By the end of the day the gear had begun to melt so we plowed it in, waited for the sun to set, and hauled back. We found a nice market cod on every other hook for a 4,500 pounds total.

Thus began a quest to fill out the trust's Gulf of Maine cod allocation, as more and more sector boats wanted to fish there. This was a typical example of a huge advantage the trust held for our boats, to anticipate our quota needs and lease them in using our strength of capital. Having a manager who knows your fleet and your fishery who can also wield capital is key.

The Nature Conservancy recently bought permits with a right of first refusal for local permit banks five years out, in several New England communities. The win-win is they have purchased in areas that are in jeopardy of losing access to their fisheries but may not have the capital to buy permits yet, **and** favoring sustainable fishing methods. I sold them two of the three permits I held, as values are currently declining and I'm "graying out."

Collaborating with Whole Foods

I am a longtime advocate of community supported fisheries and the buy local movement. However, as the cod resource became scarce in our waters, local boats were no longer able to compete with the steady-supply, fixed-price model offered by Iceland and now Norway (still called "local" fish by our restaurants).

Because of the "choppiness" of our supply, most New England fishermen sold to auctions over the past few decades. When the Gloucester Seafood Display Auction opened in 1998 I was one of its first customers. For several years we enjoyed great prices in a totally transparent, live-auction format. Then the auction moved to an electronic format where sellers were prevented from knowing the identity of the buyer. Before long we learned that variable hidden fees were being charged but not reflected on our settlement sheets.

Into this void of integrity stepped Whole Foods, a high end grocer with plans to open a Cape Cod store in 2014. With the help of the trust, we sat down with regional seafood managers and educated them about our fishery and our story. Then we set up an ex-vessel price structure and marketed our product. Whole Foods appreciated that we were a well-managed sustainable fishery. We still meet with Whole Foods regularly and consider them a partner in our business. Without the certainty of access to the resource afforded to us by the Cape Cod Fisheries Trust, we could not enter into a relationship like this.

Thank you for the opportunity to speak at this workshop.

Eric Hesse, Cape Cod fisherman. (Deborah Mercy photo.)

Financing Fishing Communities: Strengthening US Fishing Communities Initiative

Kelly Wachowicz
Strengthening US Fishing Communities Initiative, Danville, California

I have been working on the financing and restoration of fisheries for the last four years. Before that I worked for 20 years as an investor, mostly in real estate and timber and some in infrastructure. I got interested in fisheries because although I don't come from a fishing family and I haven't done much fishing except for trout in mountain streams, my family is very connected to the sea in various ways and I've always had a love for coastal communities. The more I learn about fishing communities the more I love them.

I have in recent years worked with the Cape Cod Fisheries Trust, to explore and expand the trust activities and to lend the trust ideas to other fishing communities in New England. More recently I have worked with several international fisheries with Bloomberg support. We published a number of investing strategies for impact investors; see www.investinvibrantoceans.org. There are very detailed investing strategies that highlight—from the investing perspective—what matters in supporting communities and trying to deploy capital to support fisheries. In December 2015 our team launched a new initiative that's been supported by the Walton Family Foundation: the Strengthening US Fishing Communities Initiative. Our objective is to find ways to support fishing communities that are seeking to secure fishing access and protect their resources. We are planning to spend the next six to nine months developing a plan of action and a strategy, and then get support to implement strategies in partnership with communities over the next two years.

I love hearing stories of fishing communities. Sometimes they are success stories. Sometimes they are stories of struggle. I don't want to romanticize the fishing life as an outsider, and I know every fishing community and its fishing traditions are different. But what they seem to have in common is a passion for coastal experience and a connection to the wildness of the sea. There are always stories about family and the next generation. Part of the reason I like working with fishermen and fishing communities is that I always find something that I want to bring back to my community. I am pretty much a city mouse at this time—I've lived for 20 years in New York. There is probably a lot that urban communities can learn from rural and smaller communities that have for so long demonstrated resilience, hard work, and commitment to family. From a fishing standpoint I believe and dream that there could be a future where the rest of us can learn from the example of the strength and viability of small rural communities in providing food for us all to eat and sharing a way of life.

Our initiative will be engaged in four main activities:

1. Survey community needs. I look forward to collaborating with Courtney Carothers, Rachel Donkersloot, and others who have spent a lot of time in communities in Alaska trying to understand the needs.
2. Design support mechanisms. Support mechanisms potentially are even services that would reduce the cost of what has to be done directly in the community, but could make it more possible to administer different kinds of programs with economies of scale.
3. Partner with communities.
4. Finance community needs. Financing community needs could be translated into financing quota acquisitions, which has been done with impact investors in the case of the Cape Cod Fisheries Trust, but might include other types of support as well.

The initiative is funded to support engagement with up to 20 communities nationally. We will be working with additional communities in New England, a few communities in the Gulf of Mexico, a few on the West Coast, and hopefully and significantly a number of communities here in Alaska. You'll hear more from us about the first three activities in the months to come. Here I will spend more time on the financing side of the equation, which is my primary role in developing the initiative.

I want to share a bit more about impact investing. A lot of fishing communities are wary of—and rightly so—outside money coming into the community, and what it means. There is an emerging group of investors that we call impact investors. Impact investments are made to companies, organizations, or projects with the intent of generating measurable social and environmental impacts alongside a financial return. The goals of these kinds of investors differ from more traditional investors in that they really want to align their investments with values, communities, and people. The value of impact investments in supporting these initiatives is to increase the amount of capital available to sponsor projects. We know there is a limited amount of philanthropy, and budgets in state and governments are declining or are under stress in terms of capital that can be attracted to support the strengthening of fishing communities.

There are several examples of impact investors. The Rockefeller Foundation, Packard Foundation, and Omidyar Foundation are very active. Pension funds and university endowments such as Ullico and Syracuse University have impact investors. The Blue Haven Initiative has committed its entire investment portfolio of $500 million to support impact investing. Some larger banking institutions are now developing impact investing arms, such as CSFB, Goldman Sachs, and JP Morgan. Smaller organizations and companies such as Root Capital, Bear Tooth Capital, RSF, and BlackRock have begun to match investors with values-driven objectives, to support communities and projects. And Kickstarter and Indiegogo allow you to aggregate capital from smaller investors, which also could be an interesting angle to support efforts here in Alaska.

There is a lot that could be said about the evolution in the investment community to focus investing activity on ethically and socially responsible business projects. I want to turn to why investors are excited to invest in and support fishing communities in particular. I have had the opportunity to meet with dozens of impact investors over the past four years. There is real enthusiasm for supporting fishing communities and businesses. Investors like fisheries and communities because they like being part of the food system just as a matter of good business. They like that capital can support communities and conservation all at the same time. When fish stocks are healthy and communities thrive, that means business can be profitable and it makes good sense. Investors like the idea also because they know fishing communities and people involved in food production represent some of the best that America has to offer, and they like being associated with it. Even when fishing communities are struggling they demonstrate resilience, hard work, and tradition. It might be helpful to shift our mindset from viewing the fisheries access challenge in small communities as a problem to fix, to thinking about the opportunity that exists in building and supporting economic development and investing in small fishing communities. It is an incredible opportunity to support something that is so important to our national identity and beyond.

There are also some constraints and barriers for impact investors to overcome in terms of providing capital to support fisheries. The most significant constraint is the small scale, and the expense associated with administering programs across a fragmented set of communities as well as the cost from an investing standpoint of doing due diligence to investigate and understand the opportunity profile. The cost can be very high when you have to repeat it multiple times for a small scale investment.

What would facilitate impact investment to support fishing communities? (1) Strong leadership is always key—working with leaders in businesses and communities that have integrity, are competent, have ambition, and are also realistic. (2) Local sponsorship is key. Do permit banks or community businesses have relationships in the community? Are there relationships with other local leaders, with community development banks or local NGOs that can strengthen the business strategy of the community organization or businesses? (3) Clear business plans are needed. Investors have different requirements or objectives in terms of their actual (4) measurable community impacts, or (5) conservation impacts they are trying to support. There is a real opportunity here with the initiative to (6) aggregate projects. Where small scale can be a challenge, there may be ways to structure a set of shared resources to support multiple communities in a similar fashion and reduce costs. Due diligence is part of the equation.

On the financial side, what kind of returns does an impact investor seek in order to be willing to part with capital and engage with the community? There is a wide range of expectations. Some investors support projects that might generate a 3% return, if it meets all the other social and environmental criteria they are trying to align with. But if a higher return can be offered that opens up the door to other pools of capital. Depending on the nature of the project there could be many different opportunities.

Our work with the Strengthening US Fishing Communities Initiative is aimed at helping fishing communities develop leadership and community sponsorship, and to design and structure strategies that are workable, and ultimately to deliver capital to support communities. Every investor is different. The investment approach our team would like to foster is one that is fundamentally predicated on creating partnerships with communities, a relationship that respects community leadership, and activity that is focused on equity and is based on kinship and mutual respect. Often it's the case that when money comes into the room it can distort the dynamic of a working relationship. It is our objective to find ways to partner where the partnership is a collaboration with shared problem-solving and a commitment for the long term.

Our invitation to you is to bring your best ideas. Let's work together and see what we can achieve.

Small Group Discussions and Opinions

During the workshop, Fisheries Access for Alaska—Charting the Future, small groups met to define issues and suggest solutions to the lack of access to fisheries by Alaskans. A summary of those discussions is presented here. These notes attempt to capture the dialogue of participants in the small groups and differ in style based on summary notes available from each group. The notes were not edited for accuracy, nor do they always represent consensus statements from the groups.

1. Problem statement, guidance statement, and measures of success

Problem

- Alaska's fisheries are a critical and sustainable source of employment, income, cultural identity, and state and local tax revenue for the state of Alaska and its residents. In many fishery-dependent communities there is a lack of alternative economic opportunities. There is a lack of awareness in our state regarding the value of fisheries and fishing communities and a lack of acknowledgment of the importance of mixed cash/subsistence economies in rural Alaska.

- Permits and quota are leaving communities and local residents are not able to afford to retain access to or enter fisheries. Ownership of Alaska fishery access rights by non-Alaskans has grown. Permits have migrated away from traditional fishing communities, and there is a subsequent loss of opportunity, heritage, infrastructure, and community health. Loss of access leads to the degradation of communities and consolidation. This leads to a lack of economic opportunity in fishery-dependent communities and an inability to keep the value and economic benefit of Alaska's fish resources within the communities and individuals of Alaska.

- Access to fisheries encourages the next generation to remain in the community. There is a lack of desire by youth in some communities to engage in fishing and a lack of reliable captains in the area to mentor youth due to limited permits and quota. There is no specific entry path for the next generation or exit path in our state that supports local access. There is a need to build a more positive view of fisheries among youth and recreate the pride in fisheries.

- Fishing access rights in Alaska are primarily for market-based transfer of permits and quota. The high cost of fishing rights inhibits an individual's ability to have a diversified fishing portfolio. There is no state policy to prioritize community access to fisheries resources.
- Solutions are often short-term fixes or are subsidized by short-term funds, so no fundamental change is able to happen.

Guidance statement

- Alaska's "Fish First" policy needs to include ensuring access to fisheries resources by Alaskans and residents of fishery-dependent communities. The State of Alaska must engage and renew efforts to stabilize and maintain healthy, sustainable, fishery-dependent communities.
- Strengthen local communities and their cultures, economies, and quality of life by preserving local commercial fishing opportunities.
- Develop programs and policy that will reverse the trend of outmigration of fishing from our coastal communities and our state overall. Realign the value that's going out of state so that we can create opportunity in our communities. There is a need to look for long-term solutions for the future of fishing.
- Design a means to encourage permit holders to stay in communities, and eventually reverse the outmigration of permits trend.
- Increase policy that enables fishing fleets to operate in diverse fisheries.
- Ensure future generations have fishery access; foster stewardship and community stability within present-day and future fishermen.
- Build the capacity and human infrastructure to maintain and sustain fisheries through education, mentoring, apprenticeships, and financial training, including a cultural component.
    - ◇ Create a strategic plan to increase access to Alaska fisheries by Alaska residents, with measurable goals such as:
    - ◇ Increase in percentage of permits held by younger fishermen (under 40 years).
    - ◇ Increase in number and percentage of permits held by Alaska residents.

- ◇ Increase in number of local vessels.
- ◇ Increase in number of residents engaged in multiple, diverse fisheries.
- ◇ Consider social goals.

What does success look like?

- Healthy coastal communities with clear processes for young people to enter fisheries, and a clear vision of and provisions for transition of those retiring from the fishery.
- Coastal Alaska has thriving, stable, intergenerational participation in local fisheries (generational continuity, financial stability). Communities that can support fishing businesses, with infrastructure and amenities that make people want to stay. Basic income from local fisheries is part of a diverse local economy.
- Measurable growth in the number of fishermen and vessels residing in coastal communities. Permits and quota increasingly held by residents of local communities.
- Increased interest in fishing careers from coastal participants.
- Increase in young fishermen entering fisheries with confidence they can make a viable living. More young people are able to earn a living by staying local and working as a part of their community.
- Prosperous, home-ported fleets owned and crewed by residents of Alaska's coastal communities with mechanisms in place to ensure entry by future generations from those communities. Provisions made outside of monetized fisheries access for youth and small-scale fishermen.

2. Solutions that are geographic, regional, place-based/community-based

The Bristol Bay permit loan program

Bristol Bay has seen a steep net outmigration of salmon limited-entry permits from local watershed communities. The Bristol Bay Economic Development Corporation (BBEDC), a local CDQ group, began looking for solutions to stem that outflow. Legalities and unanswered questions posed major initial challenges. BBEDC's current program to keep permits in the region has gone through many changes, and has yet to truly stem the tide of outmigration .

There are strong opinions on why other efforts to prevent outmigration of access haven't moved forward. Even with financing assistance, the high costs continue to be a barrier. A $175,000 commitment on top of gear and vessel investment, regardless of assistance, is still intimidating to many in the region—especially young entrants.

Residency and sweat equity requirements and structure to the program are essential. This incentivizes people to stay in the region and continue to work the permit. (People are eligible after a 12-month residence in the region.)

Challenges include getting young people interested and committed, due to the high requirements of time and money. They also have to qualify for a loan, and half of the applicants end up not qualifying. Lack of credit or bad credit history and intimidating costs continue to be barriers despite a robust program for financial assistance.

Market conditions—large returns, low prices, flooded markets, processor instability—create a lack of confidence in the industry. These factors may be preventing people from taking access opportunities because of both actual and perceived risk.

Question: What is the loss when they move out of the region? Answer: You have to pay BBEDC all of the money they've invested in you. You can retain the permit, but you have to pay the loan and any money BBEDC has granted to you through the program.

Halibut in Bristol Bay

A person has to be a resident to lease out community development quota (CDQ) halibut quota. The International Pacific Halibut Commission (IPHC) allocation is open to the 17 villages in the CDQ program. The season is open until the quota is caught. But the program has yet to max out the quota. Fishermen have to apply for fish quota, can't be in non-compliance with other programs, have to be a resident, and must be fishing with a resident-owned boat. But once they qualify there is not an allocation per fisherman. It is a micro-open-access fishery under the region's established allocation. This may be a good solution to look at for in-region flexibility and local access control that still fits within the allocation structure. This and other examples of modified season structures are a possible solution to local access management.

We have a salmon-centric attitude and approach in Alaska. Diversified species strategies may not gain ground in statewide conversations. It may be better to focus on salmon first, our common thread, and then solve other species issues.

What are next-step solutions?

- Permit banks could be a solution, but it matters how it is going to be managed. If the bank generates revenue through the permit lease, there are greater implications. What will the future look like under those new conditions? How will it impact the individual owners? The ability to hold it until there is a qualifying applicant to sell it to may be better than leasing.

- If we're trying to avoid the concept of leasing, then it may be better to structure it as a sale somehow. Avoid the leasing structure within the permit bank. Perhaps a nontransferable permit could be available to be used by a qualifying person in the community. A person can make revenue from that use, and then decide if they want to buy their own permit. This wouldn't necessarily be purchasing or leasing the community permit, just qualifying to use it as a means to enter into the fishery. We just need an option that is less financially onerous and intimidating than the current system.
- Carefully planned parameters and limits on percentages of permits that go into a bank could help address concerns for individual buyers and market distortion. But any removal of permits from the established market would create a factor of price increase for the remaining permits.
- Has the BBEDC program or effort affected market dynamics? Some feel that it has not.
- Is there a way to create a structure like the Maine program, where residency requirements, application programs, and waiting lists all help decide who gets the permit and how it best benefits the community?
- A way to transition might be to have more of a holding company, not a bank. A place that could facilitate the inward migration of permits. For example, in Bristol Bay a local teacher prioritized giving purchasing opportunity to a local resident. BBEDC helped facilitate that. Is there a way to make this kind of example more standardized? Giving sellers an option to direct the sale of their permit to a local resident, and an institution the ability to facilitate that?
- Rights of first refusal—the experience in the federal fishing system is that this option has significant challenges. The time frame is often impractical in terms of carrying forward the original intent.
- Note: It is important to prioritize low administrative costs for any of these solutions.
- There is a need to connect how this works for individual regions and communities, to how it works for the state overall.
- Unless a permit bank has a path to helping people on the way to ownership, we can see it not making any difference. Fishermen still

can have the same barriers of credit, etc. Benefitting the communities needs to be the bottom line, rather than the benefit to the individual. The structure needs to reflect that priority.

- Anchoring permits to the community is important, rather than relying on the dynamics of market or an assistance institution. Many like the micro-open-access concept.
- The conceptual challenge of this is only the first step; the financing for making it come to pass is very challenging as well. That plan must be developing alongside the conceptual development.
- There is concern for complexity creating barriers in the legislature. But we need the complexity to get through the constitutional challenge and have a thorough enough concept to succeed. This requires careful legislative strategy.
- Maybe the concept of permit banks really needs to be open to region-specific limitations and structures. There are so many different needs. One size will not fit all the varied regions.
- We often focus too much on status quo and working within established programs. If we look to our international examples, we see people taking small pieces of allocation out of the status quo, and creating an entirely new program. Perhaps this is how we should be looking at our options. Maybe we should be giving ourselves the capacity to try an entirely new structure on a regional, small-scale basis in order to not stay attached to the status quo and the same challenges that come with it. Maybe we don't need to focus as much on comprehensive fixes, but on smaller scale solutions that add benefit to the community.
- If this group of people, interested in changing our access pattern, wants to effect change in the current legislative climate, we need to work hard to find a unified vision and present it together. Otherwise, we have far less hope of progressing.
- Anything brought before the legislative body needs to have some kind of kill switch—a mechanism that allows us to remove something that gets transformed into something it was not intended to be.
- Note that there may be more individuals than we realize who are willing and interested in pointing their permit sale to a local buyer. But we are still in a capitalistic environment. As we look at strategies for acquiring permits or quota, we need to remember that. Sellers may be willing to participate in a positive redirection of ownership to local residents, but we would be strategic to have in place a mechanism for acquiring fair market value.

- It is possible that the skeletal structure of the permit bank is something that could be applied in multiple regions, but each region could have autonomy over the specificity of the application of its priorities.
- What's really been fleshed out is that a region-specific approach is likely necessary, considering the different tools for financing and distribution, and varying needs, by region. Also there is a need to anchor quota/permits to community. They could act as a clearinghouse/facilitator in one function, but also retain a small amount of permits or quota to provide consistent access points for residents.
- Maybe there are ways already within the permit bank bill to create this regional approach, while still maintaining a skeleton that contains the principles and values vital to original intent.

New ideas

An apprenticeship program, like the Maine lobster program, is a possibility. An expansion of the educational permit process could create permits instead of buying them out of an open market system. This could take some of the unknowns out that are created by transferability of limited entry permits.

What's great about the student track of the Maine program is the clear reward. Perhaps an expansion of our educational permits needs to have that path to ownership and independence built in—motivation to enroll in and carry out the educational program.

An educational permit wouldn't transform, but it could create a link between an educational permit/program and a financing program like the BBEDC program that facilitates both the initial opportunity and the eventual path to ownership. This could allow people to find a path to skill building, people who don't necessarily already have an in through a family member or friend.

How can we build in help with equipment and vessel financing? A permit or quota is, after all, not enough. An educational permit could possibly be used as a stepping stone (through skill building, apprenticeship, sweat equity) toward eventual ownership of a permit being held by a permit bank.

There is a strong recommendation to continue this conversation in a focus group or the steering committee that can make concrete recommendations and a clear path forward reflecting these discussions.

3. Solutions that target youth, education, and access for next/future generations

Issues

- Many Alaska communities are missing a generation of fishermen. There is a noticeable lack of work ethic among some youth.
- Fluctuations in the price of fish lead to uncertainty about the profitability of fishing.
- A negative view of fishing within communities has been documented in some cases.
- Obstacles to getting funding have been observed, such as loans denied.
- Increasingly complex regulations and technical requirements, such as safety laws, increase costs to a fishing business.
- There is a lack of a sustainable exit strategy for fishermen.
- Fishing permits are leaving Alaska communities.

Solutions

- Mentorship/internship/apprenticeship programs. A fishing vessel/family could take on a young external crewmember. Provide an incentive for that vessel/captain to take on inexperienced crew, perhaps operating in the late fishing season.
- Provide an incentive to encourage young fishermen to become involved in fishing. A certificate could be issued that gives a young person a monetary reward in the form of discounted permits or tax reduction, etc.
- Create a policy for mentorship/internship/apprenticeship.
- Educational programs within the school system that focus on and/or incorporate fishing. School programs focused on fishing—use successful models.
- Small-scale incorporation of fishing as a career option in career fairs or classes to promote recognition of fishing as a viable career and promote fishing culture.
- Educational boat/fleet to use as a learning tool and give hands-on experience to youth.

- Holistic educational opportunities for both youth and current crew. Provide off-season opportunities to take other useful classes (economics, technology, safety, credit building, etc.). College classes could be subsidized.
- Consider tapping into impact investment funds.

4. Solutions and tools that target indigenous, cultural, livelihood, and/or community rights to fisheries

Preserve the human right to fish

The UN Human Rights Council ruled that Iceland's Individual Transferable Quota (ITQ) system violated the human right to work. This led to the creation of a seasonal open-access fishery to support community development in regions with declining fisheries access.

Use educational fishing permits

An educational fishing permit could be developed for both commercial and subsistence fishermen. A permit would have to be developed within the Alaska Native Claims Settlement Act (ANCSA) structure, and would need to be consistent with state law.

In the Kenaitze educational fishery, the kids harvest the fish and they also give them to elders. It builds community and is really beautiful.

Presentations by State of Alaska employees supported the increased use of educational permits. The State's recognition of the sovereign rights of tribal governments in Alaska could help to re-establish hunting and fishing rights. Tribal sovereignty is also recognized at the federal level.

Allow multiple users of a limited-entry fishing permit

In the Shumagin area the graying of the fleet is evident. If a locally owned permit is sold, it's going outside. There are only four drift net salmon permits left in Sand Point—there were over 100, possibly 150.

Consider the idea of a permit where you are the owner—but you could add other names to it, allowing others to fill in fishing effort. There would be no change in ownership and the permit holder would not need to be onboard at all times. This allowance could also be limited to permits held by older fishermen. This mitigates the need to sell the permit as the resident owner gets older and cannot be onboard each season.

Recognize tribal sovereignty

There is international attention for the protection of fishing rights for indigenous peoples (e.g., Maori and Saami peoples). In Alaska, one step in this direction would expand subsistence fishing rights to include commercial access for rural users, recognizing commercial fishing and its contribution to subsistence, to culture, and to community viability.

Create a small-scale community-based fisheries permit (e.g., salmon beach seine)

Enabling small-scale commercial harvests near communities and traditional use areas would contribute to regaining lost fishery access in some regions. Dr. Langdon discussed this as a possibility for "forgone harvests" in Southeast Alaska. A small-scale, community-based fisheries permit might be difficult in other regions, such as Bristol Bay, but it could be done in Levelock, for example. Conservation is an important consideration.

Nondalton residents are allowed to beach seine for salmon. In Bristol Bay, once the fish get past the upper tidal limits they are off limits for the commercial fishery. For Ugashik, a community-based permit would need to be anchored to the local region. Or you could consider salmon that have passed into the river in Aleknagik as a potential "boundary."

Develop an open access pool

An open access pool for community use may be possible using community quota entities (CQEs) as established in the Gulf of Alaska to manage local access and using traditional ecological knowledge. Regional or community boundaries would need to be established, and a harvest limit for the area.

Tribal permits

Tribes must have the ability to engage in government to government relationships and consider tribal permits. There is a need for education and training for tribes to engage in tribal consultation with state and federal management agencies.

Issue permits for permit banks

When the State auctions off the permits, instead of selling them they can issue them through permit banks. They can take 1% of the stocks and add it into the pool. Halibut shares are in units, so you can take some of the pounds based on time of purchase.

Some organizations are trying to convince federal managers to charge harvesters for the poundage in an initial allocation under a catch share system in which harvesters are allocated fishing privileges or for mits under a limited access program. The federal License Limitation ogram (LLP) permits to harvest groundfish and the charter halibut ermits were issued free of charge based on qualifying years, as was halibut and sablefish individual fishing quota (IFQ).

Regionalization of landing or port of landing requirements

Regionalization of landing or port of landing requirements could serve to stabilize a community's landings in a federal fishery by requiring that harvesters deliver to a specific port or region. This protects both shoreside processors and the community dependent on those landings rom significant changes in delivery patterns. There is also an issue with ishermen having to travel to deliver their catches to the Alaska seafood market.

Community quota entities

CQEs in the Gulf of Alaska—nonprofits eligible to purchase halibut and sablefish quota—have been expanded to include charter halibut and Pacific cod permits, at no cost to the CQEs. In the western and central Gulf of Alaska the organization is there, but there is little halibut/sablefish quota due to lack of capital to purchase quota. Success in acquiring halibut and sablefish quota would achieve long-term access to a portion of the resource. Consider allowing Bering Sea/Aleutian Islands community development quota programs (CDQs) to partner and/or loan to CQEs.

Other options discussed

- CDQ for the Gulf of Alaska.
- Relax the three-year emergency transfer requirements.
- Allow elders to lease permits to family members.

5. Solutions that address access to financial capital and the pathway to get there

Financial education

Expand education in basic financial literacy and financial management to prepare individuals to be better equipped to obtain capital.

Work currently under way/current resources

- Bristol Bay Economic Development Corporation (BBEDC) has a program to link permit financing to financial management training.
- The Alaska Young Fishermen's Summit, held biennially, includes financial management and planning in the program.
- State of Alaska Commercial Fisheries Loan program.
- Alaska Sea Grant FishBiz resources/revamped website.
- Small Business Development Centers.

Issues to address

- What age and place to start? Junior high/high school.
- Can models from the Alaska Housing Finance Corporation (AHFC) be applied?
- Cultural challenges/barriers.
- Strained capacity in the educational system/schools to offer or expand such programs.
- Create the program, but will people show up? Is the program geared to fishermen or potential fishermen?
- Should the program be mandatory/be required by CDQ groups, lenders, or other entity?
- What other incentives can be used?

How to get there

Education in basic financial literacy and financial management to prepare individuals to be more equipped to obtain capital.

Apprenticeship programs/educational permits

Establish apprentice, educational permit, or junior permit programs to address the high cost of permits and facilitate entry into fisheries, and grow the pool of qualified applicants.

Work currently under way/current resources

- Establishing pathways to entry that avoid the complex financial details that can accompany fishing businesses. State loans for permits address this. They address the goal of facilitating entry, have young applicants, and a high success rate: $92 million in loans out, 90% of assets lent.
- Is there a need for stronger communication about the State program?
- Alaska Longline Fishermen's Association (ALFA) currently has an informal apprenticeship program on a Sitka troller. There is potential to expand this.
- Attempts to change the NMFS fisheries finance program and open up a Maritime Administration loan program.

Issues to address

- The State only requires a crew license and that a person fished the last three out of five years, including the year preceding the application.
- Educational permits are an underutilized tool to help young Alaskans gain fishing experience.
- Support of skippers in helping crew step up in the business and acquire quota.

Trusts/permit bank

Fisheries trusts, permit banks, or alternative arrangements may allow for sharing risk and potential to earn equity in a permit or quota over time.

Work currently under way/current resources

- Alaska Sustainable Fisheries Trust, Local Fish Fund.
- Rep. Jonathan Kreiss-Tomkins' community permit bank bill is under development.

Issues to address

- What is the actual or potential of impact investing to play a role here? There are not many good examples of success. It is likely that it needs to be linked explicitly to environmental and social outcomes.

- Integration of philanthropy and investing may provide opportunities.
- The time period for qualification to address issues or ensuring opportunity for Alaska and local community residents.
- Impacts on current permit/quota values and views of current permit/quota holders.
- Financing or access to the boat and gear once you have the permit; how to avoid issues with processors taking advantage of this.

Policy

Balance policies that result in consolidation and therefore increase the price of quota/permits with the need for financial viability in fisheries.

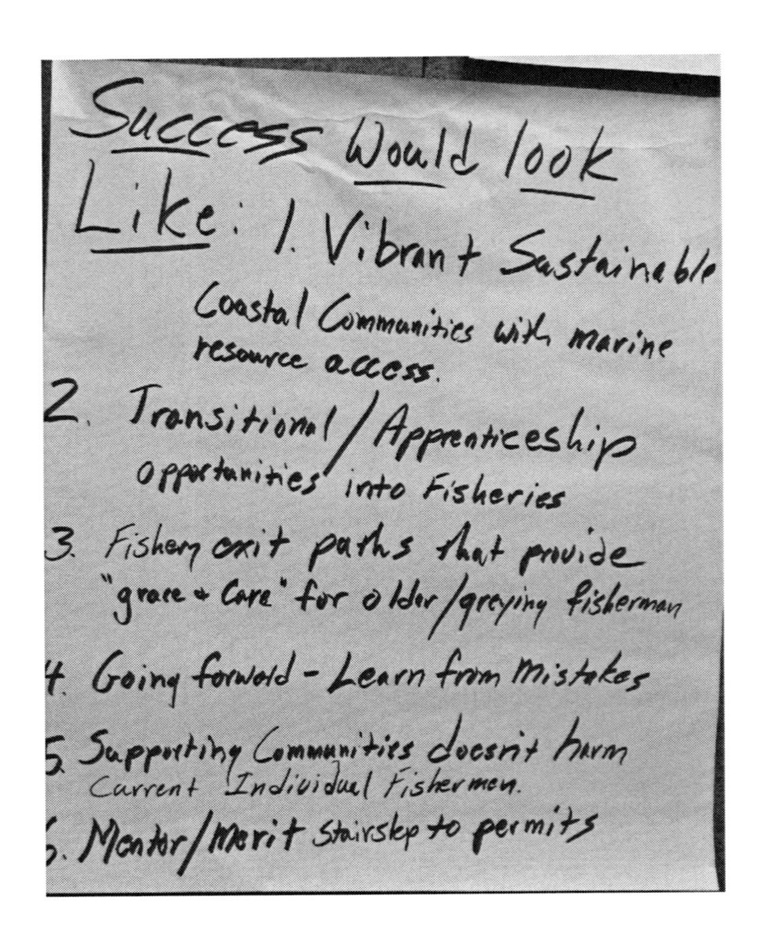